The Inspirational Untold Stories of Secondary Mathematics Teachers

The Inspirational Untold Stories of Secondary Mathematics Teachers

Edited by

Alice F. Artzt

Frances R. Curcio

INFORMATION AGE PUBLISHING, INC.
Charlotte, NC • www.infoagepub.com

Library of Congress Cataloging-In-Publication Data

The CIP data for this book can be found on the Library of Congress website (loc.gov).

Paperback: 978-1-64802-201-2
Hardcover: 978-1-64802-202-9
E-Book: 978-1-64802-203-6

CONTENTS

ENDORSEMENTS

This lovely book contains personal stories about the process of becoming a mathematics teacher and the challenges and rewards of the early years of teaching. These stories highlight that the path to teaching is often indirect, rocky, and filled with doubts. But these poignant stories are powerful because they are so honest. I wish I'd read these stories before I experienced some of the joys and challenges of my early years of teaching because they would have prepared me for the roller coaster of emotion associated with entering this complex but beautiful profession. I think these stories will be helpful when working with prospective and early career teachers.

—Randolph Philipp, Ph.D.
Professor of Mathematics Education
School of Teacher Education
San Diego State University
Immediate Past President
Association of Mathematics Teacher Educators (AMTE)

This is a book about real people and true stories; the narratives are really insightful and truly inspirational. It is not only a book that those involved in teacher preparation programs may find useful and informative to read, but also a book that could provide insights and inspiration to those who are exploring what it is like to be a teacher. The journey of each of these success stories, despite the diverse start-

ing point of each, speaks volumes of the importance of an effective teacher preparation program that not only nurtures but also provides support for the growth of
the preservice teachers. The narratives in this book are certainly a testimonial to
what we often hear–Teachers are more often made than born.

—Ngan Hoe Lee, Ph.D.
Associate Professor, Mathematics & Mathematics Education
Office of Education Research
National Institute of Education
Singapore

FOREWORD

Storytelling has the potential to reveal what is in the heart and mind of the story-teller. Throughout the ages, storytelling, both orally and in writing, has been the means to pass on traditions, cultural expectations, and social mores, and can offer life lessons by communicating and connecting with people. Recently, the power of using narratives and sharing stories (Fang et al., 2016; *Tiny Teaching Stories*, 2019), and writing letters (Rising, 2019), are examples of documenting experiences while providing advice for teachers in meaningful contexts.

The purpose of this book is to extend the existing literature by connecting personal background and teaching narratives with the innovative features of a special four-year undergraduate mathematics teacher preparation program.[1] The twelve authors are secondary school mathematics teachers who graduated from the program. The range of narratives vary in every possible way, from the reasons the authors became secondary mathematics teachers, to the number of years they have been teaching, to the experiences they have encountered while teaching, to the different roles they have taken on throughout their careers. Nevertheless, one strand permeates all of their stories—their passion for what they do and their ability to reflect on the early college experiences that contribute to their performance.

---

[1]* Detailed information about the TIME 2000 Program (i.e., Teaching Improvements through Mathematics Education) can be found at qc.cuny.edu/time2000.

Although these teachers all graduated at different times from the same teacher preparation program, their narratives will shed light on what it means to teach mathematics and the components of a secondary mathematics teacher preparation program that can contribute to their resilience and competence in teaching. What we have learned from their stories is both unique to them and generalizable to all. They are human stories that shed light on the inner lives of teachers that are rarely divulged.

If you are a preservice or an in-service teacher, as you read through these narratives, you might be informed about different aspects of teaching that you had not been aware of before that will give you added insight about teaching secondary mathematics. You also might find that the multiple feelings (e.g., aspirations, doubts, fears) you have had about teaching are shared by other teachers. If you are a teacher educator, these personal narratives might inform you about how different aspects of your program affect the professional lives of your graduates. Finally, if you are an administrator you might be informed by how unexpected and unusual backgrounds of your teachers can affect their commitment and performance in the classroom.

Given the well-documented continual shortage of highly qualified mathematics teachers, you might also be attentive to the themes that arise in these narratives that would shed light on ways that you can recruit middle or high school students to become mathematics teachers, and the ways to sustain them in a mathematics teacher preparation program, and of course, how they can be retained in teaching, once they enter the field. We expect, that as you read these very personal narratives, you will see how a very strong mathematics teacher preparation program, described in "Preparing Secondary Mathematics Teachers," has contributed to the heart of what each of the authors has written.

PREPARING SECONDARY MATHEMATICS TEACHERS

Alice F. Artzt and Frances R. Curcio

It is well documented that today there is a teacher shortage across the nation. A more dire situation exists in the area of mathematics (Walker, 2019). Furthermore, according to Scharton (2018), "In the last 20 years, teacher attrition has nearly doubled." Clearly when districts find it difficult to hire qualified mathematics teachers and retain experienced mathematics teachers, the education of students is compromised. In fact, many reports from such organizations as the National Research Council (2002), the U.S. Department of Education (2002), and the National Academy of Sciences (2006), have indicated that the security of our nation and the health of our economy is being jeopardized by the *shortage of mathematics and science teachers*. Ingersoll and Perda (2009) clarified that the demand for new math teachers is mostly a result of preretirement turnover, caused by job dissatisfaction. The documented reasons for the shortages and high attrition rate are attributed to subpar salaries, low status, limited professional support, and difficult working conditions (Mulvahill, 2019). Much needs to be done for these conditions to improve. However, as teacher educators, the best we can do is to try to prepare our preservice teachers for the challenges that lie ahead, as they start their teaching careers, knowing full well that they will be underpaid, that the teaching profession is not properly valued, yet still have a dream to help others learn mathematics. How can they be prepared? What experiences are they draw-

The Inspirational Untold Stories of Secondary Mathematics Teachers, pages xi–xvii.
Copyright © 2020 by Information Age Publishing
All rights of reproduction in any form reserved.

ing on as they meet the obstacles that present themselves in the classroom? Some of the answers to these questions are relevant to all who are in the teaching profession, as they shed light on the most personal inner thoughts and feelings of teachers who are on the "front line." Through the inspiring narratives reported in this book, we see how teachers use their personal experiences and experiences from their teacher preparation program to cope with these challenges so that they can persist as teachers in their goal to help others learn mathematics. What follows, is background information about the program from which these particular teachers graduated, called TIME 2000 (i.e., Teaching Improvements through Mathematics Education). This program has been documented in several publications (Artzt & Curcio, 2007a, 2007b, 2008; Artzt, Curcio, & Sultan, 2013; Curcio & Artzt, with Porter, 2005). We describe it here so that the reader will understand the context from which the authors speak and possibly give the reader ideas for elements of mathematics teacher preparation programs that are possibly essential for the recruitment and retention of high-quality mathematics teachers.

THE UNDERGRADUATE MATHEMATICS
TEACHER PREPARATION PROGRAM

TIME 2000 was created in 1997, and funded by the National Science Foundation as a two-year program designed to recruit mathematically talented high school seniors into a mathematics teacher preparation program that would provide free tuition for their first year of the program. The first year of funding was for coursework planning and recruitment and the second year was for student tuition and courses and seminars for students. As stated in the original grant proposal the goals of the program were and still are to (1) recruit high school seniors who do well in mathematics into a secondary mathematics teacher preparation program that begins in their freshman year, and (2) prepare them to become highly competent secondary mathematics teachers and leaders who love and understand the importance of mathematics and who are both knowledgeable and passionate about what they teach and how to teach. By securing other government grants and private funding we were able to capitalize on the accomplishments of the original project by extending it to a full four-year program and offering it to freshmen each year since 1998. After the first ten years of the program, with rising tuition costs, the grant funding was reduced from paying students' full tuition to offering them $1000 per semester for 8 semesters. By June 2019, over 250 students have graduated and are full-time mathematics teachers in the New York metropolitan and surrounding areas. As of September 2019, more than 80 undergraduates were in the pipeline, preparing to become secondary mathematics teachers.

RECRUITMENT

The program can best be described as a close-knit learning community in which participants learn about mathematics and the learning and teaching of mathematics from experiences both on and off the Queens College campus. Recruitment into the program begins while students are in high school. Students who do well in their high school mathematics courses can find out about the program in many ways. They might be encouraged to join TIME 2000 by their mathematics teachers, their school counselors, by seeing fliers in their schools, from the internet, getting letters from the college inviting them to apply (based on their high SAT scores in mathematics), or by attending a TIME 2000 event held at Queens College called, "TIME 2000 Celebrates Mathematics Teaching" (Artzt & Curcio, with Weinman, 2005). If high school students are fortunate enough to be taken to this event they are exposed to a motivational keynote speaker and then two workshops where TIME 2000 graduates, or other exciting mathematics teachers present their favorite mathematics lessons (usually involving hands-on materials, real-life situations, or technology), and then they hear a panel discussion where TIME 2000 members and graduates speak about the program (see the chapter by Leon Chu).

COURSEWORK AND FIELDWORK

Once in the program students are block scheduled into their mathematics and education courses, which are largely designed and taught by the very best professors who understand that their mission is to prepare highly competent mathematics teachers. Because the students take these courses with members of their own cohorts for four years, they form close friendships, and study groups, which increase the probability that they will successfully complete their coursework (see the chapters by London and Penagos). Furthermore, the professors collaborate in the sequencing and design of the courses, often team teach, and share grades so that they are all aware of which students might need extra support. The students are given a four-year plan that clearly outlines the mathematics and education courses they will take and when they will take them (TIME 2000 Course of Study). Aside from the courses listed on this plan, the students choose their own core courses.

A very unique feature of the program is that students take a course, "The Psychology of Learning Mathematics," during their first semester as freshmen. This is in contrast to non-TIME 2000 students, who are not permitted to take any education classes until their sophomore year, after they have completed a minimum of two semesters of calculus. The fieldwork for this course is modeled after the Japanese-Lesson Study paradigm, in that the entire freshman class along with the college professor go together and visit one master middle school teacher for three periods. Each visit focuses on one psychological aspect of instruction (e.g., motivation, behaviorism). For the first period the master teacher discusses her assignment and how she intends to address those aspects of instruction with her class. For example, she might say that she plans to motivate specific students by giving them stickers when they put their work on the board. The second period, the TIME 2000 students move to the back of the room and then view the actual lesson and sometimes even interact with and help the middle school students when they work in groups. During the third period, after the middle school students move to their next class, the TIME 2000 students debrief with the master teacher to discuss the degree of effectiveness of the psychological strategies used during the lesson. A follow-up discussion occurs at the college the following day. This experience allows the TIME 2000 students to get a view into the mind of the master teacher and see all that is involved behind the scenes when teaching a class. It also gives the college students a feeling early on if they enjoy working with middle school students (see chapters by De Sousa, and London). This unique fieldwork experience occurs again during the TIME 2000 students' sophomore year when they take a course in the "Foundations of Education" and they get to learn about the mission of various schools and how the teacher supports the mission through her daily instruction. They also get to meet other "players" who are involved in the instruction and support of students (e.g., the principal, counselors, parents, etc.). This innovative fieldwork has been highlighted in Curcio and Artzt, with Porter (2005).

Another unique feature of the program is a course that was designed for students to take the semester before they take their methods course in their senior year. The course is called "Mathematical Foundations of the Secondary School Curriculum." The purpose of this writing-intensive capstone course is to engage students in experiences that will enable them to gain a deeper understanding and enjoyment of the mathematics they will be teaching. They learn and implement student-centered pedagogical approaches, techniques of discourse, and methods of assessment and evaluation. Towards this end, students work in small groups to investigate the major content strands of the secondary school curriculum: Number and Operations, Algebra, Measurement, Geometry, and Data Analysis and Probability. They study these branches of mathematics from several perspectives. They examine the historical development of these topics and share activities that best highlight these ideas in middle and high school. They also examine and share the higher-level ideas that underlie the concepts. During the course of the semester they create and are engaged in many problem-solving activities that incorporate the use of manipulatives, computers, iPads, graphing calculators, calculator-based laboratories, and other technologies. Under the guidance of the professors they are responsible for implementing five lessons, creating and administering four related homework assignments, and one quiz. They assess their own instruction by assessing their peers' understanding as demonstrated by their classroom discourse, their journal comments, and their performance on the homework assignments and quizzes (see the chapters by Leon Chu, and London). This innovative course was highlighted in Artzt, Sultan, Curcio, and Gurl (2012).

TAKING ON LEADERSHIP ROLES

An important opportunity afforded by the program is participating in a tutoring club called, *Today's Tutors Tomorrow's Math Teachers,* in which TIME 2000 students tutor local middle and high school students in mathematics for nominal fees. The students take on leadership roles as officers of the club and take full responsibility for the tutoring club's operation. To raise money for the club they have bake sales, movie nights, and game nights, as well as run a session on best ways of tutoring. By the time students graduate, those who have tutored (which is the majority of the TIME 2000 students) have a good grasp of the mathematics curriculum and a strong sense of the typical errors students make when doing mathematics (see chapters by Leon Chu, and Payen).

Students in TIME 2000 usually have multiple talents upon which the program likes to capitalize. For example, *Reflections of TIME* is a newsletter published through the TIME 2000 Program three times per year.[1] Students volunteer to be co-editors and authors for the newsletter and get experience in using publishing software, interviewing people, creating ideas for articles, and encouraging people to write articles. The students take great pride in these publications, as they are always published before big events where the readers go beyond the TIME 2000 family (e.g., at the TIME 2000 annual conference and other events, the TIME 2000 graduation). These are experiences the students include on their résumés, as schools always have newsletters and need teachers to be moderators (see the chapter by Markinson).

An important event the students attend and help run each year is the one that is held at the college entitled *TIME 2000 Celebrates Mathematics Teaching*, described previously. The idea of this conference is to have high school juniors and seniors who are interested in mathematics, come to the college and experience mathematics in a way that they have unlikely experienced it before. Although this has actually resulted in a good recruitment effort, the experience for the TIME 2000 students participating in the conference is unique and important as their development as teachers. For example, some of them are ushers who escort the high school students to their sessions. Others serve as assistants to the presenters, making sure they have all of the supplies and materials they need for their lessons. Others serve as technology supporters who make sure any presenter who is using technology has everything working properly. Still others make sure that all of the high school students submit their evaluations of each session and then distribute lunch to them. And, others serve on a panel to share their favorite features of the TIME 2000 Program and field any questions the high school students might have. After every conference, the undergraduates share how grown up they felt "running" this conference and how much they learned from the experience (see the chapter by Kimyagarov).

FEELING PART OF A PROFESSIONAL COMMUNITY

All four TIME 2000 cohorts get together once per month when they are required to meet at monthly seminars with each other and with project faculty and staff.

[1] To view the newsletters, visit https://www.qc.cuny.edu/Academics/Honors/Time2000/Pages/Newsletter.aspx.

Presentations are given by experienced mathematics teachers, exciting mathematicians, and peers who have created technologically advanced, student-centered lessons (see the chapter by Askins). They also receive peer and faculty advisement. Once per year, graduates of TIME 2000 give a panel discussion on the challenges and rewards of teaching. The undergraduates get the perspectives of teachers, who a few short years ago were sitting in their seats as undergraduates, who have been teaching from 6 months to 16 years (some of whom are now administrators).

To help acquaint students with the community of mathematics educators outside of the college, TIME 2000 students are required to attend a minimum of one professional mathematics education conference per year. Every January, a local State University of New York college holds a conference for mathematics teachers that the TIME 2000 students attend. Usually the keynote speaker is the president of the National Council of Teachers of Mathematics and the workshops are conducted by local and nonlocal mathematics teachers, college mathematicians, and college mathematics educators. The students are always surprised that teachers attend these conferences in hopes of improving their practice. They also enjoy meeting the other attendees, some of whom were their very own teachers in middle or high school. They are particularly inspired when they see that graduates of the TIME 2000 Program are speakers (see the chapter by Markinson).

TIME FOR REFLECTION

Throughout the program the students keep in close contact with one another and with their professors. For example, they have small-group sessions, with about 8–10 students once per semester with the program director to discuss any issues that are on their minds. Additionally, at least once per semester, they write journals where they are asked to describe the courses they are taking, their feelings about those courses, what outside activities they are engaged in, suggestions they have to improve the program, and anything else that is on their minds (see the chapter by Kim).

At the end of each of their first three years of the program, they are asked to create their personal TIME 2000 portfolio. In it they are encouraged to think about how their knowledge and beliefs about mathematics, the curriculum, students, and teaching have changed since they first started. They are asked to discuss their goals for students and examine how those goals might have changed over the course of time. They are asked to support their claims with specific documents from courses or presentations they have attended, or experiences they have had. This is a way for them to keep track of all of the valuable instructional materials they have accumulated in the workshops, events, conferences and courses they have attended. Not only do students report using these materials in their teaching, but they usually bring these portfolios to interviews and more importantly, graduates have reported that when asked to compile a portfolio for their tenure review, they feel experienced with the process (see the chapters by Askins, and Ezratty).

CONCLUDING REMARKS ABOUT THE PROGRAM

The comprehensive nature of this program that has been documented above was developed over the course of many years. Throughout the program, the program faculty, staff, and students have regular meetings to assess the value of the different components of the program and through this process the program has evolved into what it is today. The closeness developed within what is now called the TIME 2000 family is what has contributed to its success, as graduates of the program always feel connected to one another, other cohorts that have graduated before them, and to the TIME 2000 faculty and staff. This support system is the reason that the retention record of TIME 2000 students in teaching after five years is over 92%. We could have asked any one of our 250 graduates to write their untold stories for this book, but we chose to focus on only twelve that would highlight the specific value of some of the aspects of the program that we have described above. The stories are honest, touching, and informative on many levels. Whether you are a preservice teacher, a teacher, or teacher educator, we believe these stories will be enlightening and, in the end, uplifting.

CHAPTER 1

MORE THAN A TEACHER

Maria Leon Chu

"What do you want to be later in life?"

"A writer," I responded resoundingly without hesitation. At the time that my eighth-grade social studies teacher, Mr. R, asked me this question, I was reading avidly and could think of no better profession than coming up with stories that would open new worlds to the imagination.

"Well, Maria, I'm sure you will become whatever you set your mind to."

I was moved at the time because this teacher thought so highly of me that he would make such a comment in the middle of class, seemingly without reason. Perhaps Mr. R had sensed in me a fighting spirit that pushed me not to fail in anything I tried to do, especially in academics. Later that year, he wrote me a letter telling me that if I ever needed a recommendation, I could write to him at his personal address. One could imagine his surprise when years later I wrote to him not asking for a recommendation, but instead, telling him that I accomplished what I had set my mind to: Instead of being a writer, I had become a mathematics teacher.

The Inspirational Untold Stories of Secondary Mathematics Teachers, pages 1–11.
Copyright © 2020 by Information Age Publishing

Why the sudden change in prospective career choice? Part of it had to do with my high school geometry teacher, Mr. B. He was charismatic, knowledgeable, artistic, and downright humorous. Throughout high school, I was very shy. In Mr. B's class though, I was a different person. At the beginning of the term, he had to coax me out of my shell, but by the second month, I was a consistent participant in class discussions, volunteering my responses not once but many times in a single session. I was proud and confident, and I thought I finally had something to contribute and I didn't have to worry about being tongue tied like I was in my other classes. Having immigrated to the United States when I was eight years old, I never felt comfortable expressing myself in a room full of peers that I deemed so much more eloquent than myself. However, geometry was something that made sense to me and I took comfort in the fact that there was usually one right answer. Unfortunately, that assurance that I felt in Mr. B's class never spilled over to my other classes and I remained shy for the rest of my high school years. However, the impact Mr. B had on me would last a long, long time.

Fast forward to the end of junior year in high school when Mr. B announced to us some devastating news. He was going to leave our school to teach at a Yeshiva. I was counting on him to teach me AP Calculus the following year, but instead we got a newly hired, young, inexperienced teacher who had never taught AP Calculus before. I was apprehensive and worried but after the first month, it was clear that Ms. L was not only knowledgeable but also very approachable. She was very upfront about not having taught the subject before, but we could all tell she was trying her best and we appreciated it. She was clear, caring, enthusiastic, and always wore a smile on her face. It was obvious that she enjoyed teaching and her joy was contagious.

Math had always been my favorite subject in school, but I hadn't seriously considered *teaching* it as a profession until Ms. L took us on a trip to Queens College. The TIME2000 program, from which Ms. L. had graduated, was holding an annual event to celebrate mathematics teaching and learning. I did not know what to expect, but it sounded like fun, so I thought, why not check it out?

On that crisp November day, there were only a handful of students attending from my school but when we got to the LeFrak Music Hall, we could feel the air buzzing with excitement. There was a huge crowd of high school students and their teachers from many different schools. A bunch of smiling college students in blue shirts guided us to where we needed to go, checked us in, and directed us to breakfast. As we navigated our way through the crowd, Ms. L suddenly paused to embrace a charismatic lady we later found out to be the program director. After a few warm exchanges, this professor turned her attention to us, the students, and asked, "And how many of you want to become math teachers?" When only one of us (not me) raised her hand, she addressed the rest of us with a glimmer in her eye, "Well, maybe you'll change your mind after today."

How right she was! The day started off with a keynote speech by Cathy Seeley, at the time, President of the National Council of Teachers of Mathematics.

She described her experiences teaching with the Peace Corps in Africa and then proceeded to tell us that education was *the* most important profession. I remember leaving the auditorium that day feeling not only impressed by Ms. Seeley's accomplishments but also inspired. Maybe one day, I, too, could make a positive difference, however small, on someone's life. After the keynote presentation, we broke out into smaller sessions. I was mind blown by the session I attended. We proved that on a grid, no matter which space we started out in, everybody had to end up in the same space at the end of a series of moves. I enjoyed mathematical puzzles, but this was something totally new to me. In a way, this was perhaps my first experience with mathematics that did not involve learning and applying it to achieve a perfect score on a test, and it felt wonderful. To discover this awesome mathematical fact just for its own sake was amazing and I was completely and utterly hooked. If I could help others experience mathematics in this exciting way, then count me in!

When I went home that day, I concluded that I must apply to this program and see where it might take me. I envisioned myself as a Mr. B, with his charisma, a Ms. L, with her love for teaching, and wishfully, even a Ms. Seeley, making an impact on the lives of those children most underserved.

With the same fighting spirit that I had in the eighth grade and throughout high school, I entered the math-teaching program in the fall of 2006. I enjoyed and excelled in my mathematics and education courses. Through exposure to the instructional methods that my professors instilled in us, I finally understood why Ms. L, despite being inexperienced and teaching a new course at a new school, was completely unfazed. In our education courses, I not only learned about co-operative learning, motivation, discovery learning, classroom management and more, but I got to experience it vicariously through our fieldwork experiences that started since freshman year, and personally as a "student" applying the methods we learned to our own courses. As we discussed hypothetical situations in the classroom, we were prepared to think deeply about how to present new concepts to students, and learned the intricacies of developing a great lesson. More and more, I felt ready to be a teacher.

In my college days, I was still shy. As I got closer to graduation, I started to wonder whether this would impede my ability to stand in front of a classroom full of teenagers and survive, let alone deliver an entire lesson without feeling the need to *run away*. Little did I know that everything I was doing as part of the program was also preparing me to overcome this hurdle.

Within the program, students organized and operated an official tutoring club which matched potential tutors (also students of the program) with middle and high school students who needed tutoring. Though at first, I was nervous that I wasn't knowledgeable enough, I quickly came to learn that the school students appreciated any help they could get and were less judgmental than I thought they would be. I tried different strategies to motivate them, including bringing proof packets with statement-and-reason-cut-outs for one student to work with as a puzzle, and even playing football with another student—every math question he answered correctly would gain him yardage while every word I spelled correctly from his dictionary would advance my pawn. As our relationships grew, I came to "fear" students less. In hindsight, it was these tutoring sessions that made me more confident in myself. They were a low-stakes way to try new things and I was rewarded with deeper student understanding.

Tutoring helped me familiarize myself with the mathematics and prepared me for the challenge in Mathematical Foundations of the Secondary School Curriculum, a course in which we taught a high school unit as a group to our fellow classmates. I was assigned to teach geometry with four of my classmates. Planning everything was the fun and easy part—We got to research different topics in geometry and then come up with cool scenarios and activities to teach them. We were to teach five lessons in total and then design an assessment to evaluate our peers' learning.

On the first day that we were actually to teach, I was scared. This was my first official time in front of a classroom, delivering a lesson from start to finish! *What if I forgot the next step? What if I messed up the math? What if I got tongue tied and couldn't come up with the right words?* Thankfully, my group had gone over our parts before the lesson and everything went smoothly. We introduced the different geometric transformations by having the "students" play with tangram puzzles and then eliciting the motions they used. We summarized them in a chart at the board and had a discussion about the properties of each transformation. Even though I got stuck a few times, not knowing how to continue the discussion, my group mates were there to rescue me. When we debriefed with our professor after the lesson, she was impressed with how well-planned and structured the lesson was. I was relieved she couldn't tell how nervous we all felt!

Our lessons proceeded and we learned a lot about making mistakes and how to fix them for the next session. This class taught me the importance of always reflecting on my teaching: What could have gone differently? What student misconceptions did I not anticipate? What would make it run more smoothly next time? Even to this day, I revise and re-revise my lessons from year to year and sometimes, even from period to period. In teaching, one size definitely does *not* fit all and the more care and time I put into reflecting on my practice, the more I can improve as a teacher. The class also provided me with more confidence about myself as a teacher and I went into student teaching ready to stand in front of a classroom full of teenagers. I am happy to report that I did not run away after all and got through every lesson from start to finish.

In May 2010, I had survived student teaching, graduated with honors, enrolled in graduate school in the Secondary Mathematics Education Program at Queens College, and found myself without a job. The New York City Department of Education had issued a hiring freeze the previous year and it had not been lifted. Fortunately, I had found part-time work at a community college teaching mathematics to adults who were preparing to take an exam for the General Equivalency

Degree (GED) in order to obtain their high school equivalency diploma. Although it was a good experience because I was able to experiment with new strategies I had learned previously, I still felt that this was not where I was meant to be. Instructing adults and teaching teenagers are very different. I just didn't feel the same sense of fulfillment that I felt during my student-teaching days.

A year later, I was scheduled to teach another GED course and a College Algebra course in the fall semester of 2011. At this point, I had not given up on teaching in a secondary school setting and had applied to several charter schools. The hiring freeze had still not been lifted and there weren't many schools that were hiring. So, I was surprised when I received an e-mail from a New Visions network school asking me to go in for an interview. I let them know that I was away, and they were willing to accommodate my schedule, setting the interview for the week after I returned.

Come the day of the interview, I was nervous, but I kept telling myself, *if this doesn't work out, I still have my job at the community college. Just breathe and be yourself. You got this!* Besides, this was a bit of a gamble anyway—it was a long commute consisting of an hour-long bus and train ride to the South Bronx, and I had heard stories about it being a tough neighborhood. But the school report card gave the small school of around 400 students an "A" for school environment and the school's mission aligned with my own: to prepare students for life by supporting their social, emotional, and academic growth. In the google search I did, the students seemed happy to be at their school and I could see myself as part of that happy family. *If the interview goes well, I'll do my best. If it flops, I have a backup and it gives me more interview experience. I have nothing to lose.*

So, I climbed the stairs to where the principal and assistant principal were waiting for me. They greeted me with smiles and told me about their school. They were impressed that although I am Asian, I speak Spanish fluently which would be great because a large portion of the students were Spanish-speaking immigrants. We discussed my credentials further and they asked me to go home and wait; I would find out their decision in about a week. I thanked them for their time and on my way out, I slowly took in the school, the ways its floors were checkered, all the positive posters hanging on the walls, the shiny red handrails. *This could be my home!* I had a good feeling about the interview, but I wasn't sure; there's really never a way of telling one hundred percent, so I prepared myself to wait out the week.

The news came sooner than expected. I was barely a block away from the school when my cell phone rang. I picked up. It was the principal. Maybe I had forgotten something? No…he was telling me that if I wanted the job, I was hired! I couldn't believe it. They actually hired me! I didn't even have to think about it; of course, I said yes! I quickly quit my job at the community college and started planning for my upcoming classes. I was to teach all ninth-grade algebra and school would start in less than a week! I had so much to do and so little time!

The week before school started, all new teachers were to participate in a professional development meeting. I was excited to meet new teachers like myself, until I realized they weren't like me at all. These other teachers had been teaching for over ten years and they were transferring from other schools. Like myself, most of them were scheduled to teach ninth grade. I was starting to feel very nervous because I really was the only newbie. *What if I messed up? All these other teachers seemed like they knew what they were doing.* Thankfully, I was assigned a mentor who had been teaching algebra to the combined ninth and tenth graders at this school for a long time. Mr. V would be guiding me throughout the year, and he graciously offered to share all his resources with me.

At the faculty meeting the day before school started, I found out that the mathematics department consisted of three other teachers, all of whom had taught at this school before and none of whom would be teaching my course. However, they were friendly and offered their help and encouragement. It was also here that I found out several facts that if I had been more experienced, would have raised red flags: The principal was starting his first year not only at this school, but as a principal; there was no math assistant principal but we would get one soon; the algebra course which was traditionally taught in two years was to be crammed into one year (I would be teaching this course); I was going to teach one co-teaching class but no team teacher had been hired yet; and about one-third of last year's teachers had quit.

At the time, I was naïve and didn't think much of these things. As Mr. R had known all those years ago, I was not a quitter and I would just try my best to overcome any obstacle that came my way. The funny thing is, I did not even perceive these things as obstacles. I was more worried about my lesson for tomorrow, how I would present myself to my students, not remembering all their names, not running away when I had the chance (I was going to be in charge of a roomful of teenagers all by myself and I'm only 5'2"!). It didn't help that at the end of the faculty meeting, a few teachers came over to wish me luck and tell me not to smile until December or the kids would walk all over me. *Why me? Was my deer-in-headlights look really that obvious? What if I couldn't fool the kids? What if I couldn't help but smile? How could I not smile and still impart my passion for math?*

The year started off well. On the first day, to get the new freshmen acclimated to the school, we did ice-breaker activities and discussed how the brain is like a muscle, which must be trained for it to develop and grow. The students were well-behaved and eager to participate. I was comforted to find that they were nervous like me about moving on to high school and I felt we could work and grow together. After the first day, I was excited to get home and start planning my lessons for the next week. I would find super cool applets, games, and real-life scenarios to engage my students. I promised myself I would learn all their names by the end of the week. But right when I was about to leave, the biology teacher walked into my room and asked how my day was. I told her about my students, how adorable they were, and how optimistic I was about our year together. "Oh sweetie, that's just the honeymoon period. Give them a week. They'll start showing their true colors." I thanked her for her advice and took it with a grain of salt. I believed that if I was firm but engaging and caring, the students would sense my sincerity and cooperate with me.

Everything went well during the first weeks. I met my co-teacher who had been teaching in a District 75 (i.e., special education) school for many years and had many classroom management techniques under his belt. Although his content knowledge was limited, he had a way of speaking to students that made them listen and do his bidding, and I was happy that I could learn from him. I thought I had good rapport with my students, too. But the science teacher's ominous prophecy came true after all. After the first test, I could see that the students were mathematically very weak. Although they had understood the lesson and actively participated during class, they still did poorly. During the test, some of them even asked me for answers to questions. It was clear I had my work cut out for me if I were to get them to perform at grade level.

It was also at this time that the behavior issues started. Calling out inappropriate things was a common issue. Sometimes, when one of my mentors (I had a coach, Ms. J, as well as Mr. V, my original mentor) was in the room, students would ask why I needed another teacher there…was it to help me? One girl challenged my authority by popping her bubble gum loudly while I was teaching. Another refused to do her work and when I approached her, she cursed at me and left the classroom. When the dean found her and brought her back, she just rolled her eyes unapologetically and put her head down. Upon seeing this, the dean asked me to give her work and he took her to his office. The students were afraid of the deans, but they viewed me as incapable any time I called the dean's office regarding inappropriate behavior. Attendance became another issue; I had about 70% attendance consistently, but it was never the same 70%. Whatever gains I made in a single day would regress during the next two days when students just didn't show up.

Because I had been exposed to numerous techniques for different situations in my undergraduate program, I employed a variety of strategies to train students to be productive. My mentor told me it's important to try new things, but I shouldn't abandon one routine because it failed a few times. It usually takes students five weeks to learn a new routine well! To encourage students to come in, settle down and start work right away, I gave out warm-up slips at the door and collected them. I raided the supply room for composition notebooks to provide students with a journal to express themselves and to keep all their notes together, as most of them tended to lose things. I incorporated learning surveys to find out about their interests and integrated my findings into the lessons. I created cooperative learning groups to encourage discussion. I used a reward system with "math bucks" to promote good behavior. I drafted contracts with individual students who demonstrated the worst behaviors to hold them accountable for their actions and teach them responsibility. I opened up my classroom during lunch time, so students could come and get tutoring or just hang out. I even started a competition between classes: the most well-behaved one would get a pizza party at the end of each month. I called parents almost daily to get students to come to class.

Some days, it felt like I had small victories: When a struggling student finally passed his first test; when we had a field day and I related to my students in a setting outside of school by working together on a project; when I received good ratings on my first official observation; when my students were excited about seeing Super Mario and Bowser when plotting inequalities; when a parent thanked me for helping her son and giving him second chances; when I assigned a project to decorate my walls to a truant student who had been arrested for graffiting and had missed months of school. I was moved to tears when he told me he didn't like certain teachers because they judged him, but he felt welcome in the few days he was in my class.

That year I learned that unlike the movies, there isn't just one magical moment where all things align, and everything goes smoothly from then on. My small victories, compared to the turmoil I faced, were just that, very small. Maybe I should have given myself more credit given the circumstances, but I am a bit of a perfectionist and most of the time, it just seemed my efforts didn't surmount to anything at all.

Managing the students and ensuring they were learning was not the most difficult aspect of my job. In November, not even three months after school started, we were called in to a faculty meeting. We received the shocking news that our school was labeled a "transformation school," meaning that it was one of the consistently lowest performing schools in the city. In June, all teachers would have to reapply for rehiring and only a small percentage of them would be able to stay. Otherwise, we would be placed in the Absent Teacher Reserve, a title that labels teachers as "incompetent," whether or not it's justified. Right after the news, we all had to put on a fake bravado to teach our students for the rest of the day. Of course, the news leaked out and students soon found out about it. They were indecisive as to whether they should stay or transfer to another school. The entire mood of the school turned pessimistic.

It was soon clear, too, that the new principal was not the ideal leader in this situation. Whereas the old principal knew all the students on a first-name basis and made it his mission to support them emotionally and socially, the new principal would often stay in his office, ignorant of the daily happenings of the school. As an effort to increase opportunities for students to gain credits necessary for graduation, the administration implemented a trimester system. It was obvious this whole thing was an experiment and we were all caught in the middle, powerless to do anything. Later in November, the tenth-grade English teacher told us about her plight during a teacher team meeting. A student had accused her of using inappropriate language and had brought her parents into the school. The teacher had denied the charges but instead of investigating thoroughly, the principal had sided with the student and removed the teacher from the classroom. Although the proceedings took a few weeks, in the end, the teacher was forced to resign, and she was banned from teaching in a secondary setting in the public-school system. This teacher had been teaching for over ten years. This incident left me wondering, *what chance do I stand, as a fresh new teacher? I am really on my own.*

It was around this time that I started questioning my decision to become a teacher. On certain days that were especially bad, I would lock my classroom door and just break down in tears. I was prepared to teach, but I never knew *being* a teacher could be this difficult. I felt that I was responsible for protecting my students against the whims of the administration and for providing them with a safe haven to learn and grow, but how could I when I couldn't even take care of myself? I fell into a spell of depression and started looking for ways out.

I applied to a high school for new immigrant students that was close to home and miraculously, I was granted an interview and demo lesson. I was offered the job even though I told them my situation about currently working at another school. The principal was very understanding but told me I needed to get my current principal's permission to leave. I was worried he would be upset and might hold a grudge against me. *What if he didn't let me leave and then proceeded to rate me unfavorably? What if as a result, other schools wouldn't take me?* Still, I mustered up courage and met with him the following day. When he told me, I was a gem and he couldn't find a replacement for me in the middle of the year, the small amount of hope left was shattered. I left his office devastated. I informed the other principal of the news and she wished me luck. Hopeless though the situation was, I knew that if I quit now, I would have to start all over but at this point, I really wasn't sure whether I was cut out to be a teacher. *How long could I last like this?*

In January 2012, I started my graduate advanced methods class at Queens College. My professor was very supportive and checked in with me from class to class. I couldn't bring myself to tell her how miserable I was, how I was losing weight fast, how I was seriously contemplating quitting teaching as a profession. I was feeling torn in so many directions. I had three mentors and a co-teacher and at this point, all their advice seemed contradictory. All the professional development I attended also appeared futile. One time, when I was assigned to plan a curriculum with one of my mentors and a substitute was covering my class, a fight broke out between two students. When I returned at the end of the period, there was blood on the Smartboard, on the floor, and on my desk. My students were horrified. I had to reassure them the next day and make sure that they were emotionally stable after witnessing such violence. If I had been there that period, I would have known my students enough to prevent the victim from taunting the attacker to a breaking point. If I had been there, maybe I could have prevented the fight. But instead, the administration had scheduled me to plan curriculum that wouldn't even be implemented until the following year, when I wouldn't be working at the school anymore. I was frustrated at how powerless I was.

In all this, the other teachers gave me advice and encouragement, opened their doors to let me cry freely in their room. The social worker even gave me advice for days when students misbehaved. Making light of the situation, she told me to imagine a pigeon flying in through the window and defecating on the students' heads! I knew that my colleagues were all trying their best to help me, but I just couldn't understand why I was so miserable or how to turn it around and go back to being the fighter that I had always been.

Finally, after an observation, my new assistant principal of mathematics and science shed some light on the situation. She saw what a hard time I was having coping and pointed out, "Failure is new to you, isn't it? I bet you always did great in school and this feeling of being out of control is new and strange. But you shouldn't be so hard on yourself. Don't take things too personally. The kids like you or else they would have run you out by now. You're actually doing fine for a new teacher handling freshmen." She was right. I wanted everything to turn out fine. I wanted to be Mr. B and Ms. L and Ms. Seeley. I wanted to make a difference and all I could think about was how much better off the students would have been with a more experienced teacher. I needed to take baby steps and forgive myself for the mistakes I thought I had made. I had taken on too much blame.

One thing that kept me going was attending my advanced methods course. I got to experience teaching mathematics without all the pressure and negativity. I got to listen to other teachers share their struggles and was comforted by their stories, some of which were similar to mine. A close friend of mine, Mr. C, who was in my same cohort in my undergraduate program, was also taking this class. It was also his first year at a public high school in Queens, and he listened to my woes and gave me advice. He told me about different strategies tried by teachers in his department and he encouraged me to hang in there. My professor also constantly showed her concern for me and tried to comfort me. I don't think I would have made it without this network that was built during my time in the program. I trusted them and they believed in me. Their support gave me the strength to finish the year.

I did not reapply to my school. Instead, my friend, Mr. C, told me about an opening at his school. The hiring freeze had been lifted only a few days after I took the job at my current school and he started the school year when I did. I interviewed with the assistant principal of the mathematics department at this high school in Queens and was given the job. I could not believe my luck! I would have a fresh start!

This is my eighth year teaching at this school that houses over four thousand students. I no longer have my own classroom, but that's a small trade-off for all the things I gained. Most of the teachers in my department graduated from the TIME 2000 Program and I feel as though I fit right in. The assistant principal is supportive and gives great advice. He always reminds us, "The math you teach your students is not most important. Everything else you do for them is. You don't teach math. You teach students." Although there is still political drama ranging from administration changes to new curriculum through the introduction of the Common Core Learning Standards to changes in teacher evaluation, I still feel at home in my new school. Like a family, we share our complaints and present a united front. We strive never to compromise our students' best interests for some hidden agenda despite any changes that come our way. Some students still come in with low math skills, others with attitude and all that encompasses being a teenager. These things are easier to deal with now.

Once in a while, students ask me why I became a math teacher, with wonder, skepticism, admiration, and once, even repulsion, in their tone. My go-to answer is that math was my favorite subject and I had an inspirational teacher. And then the other questions they like to ask are: Isn't teaching hard? Do you regret it? Looking back, I almost answer affirmatively but then I wouldn't be completely honest. On some days, teaching is like climbing a vertical hill, it just seems that impossible. Then again, when I think back to everything teaching has done for me, I can't say I regret it. Speaking in front of a classroom has done wonders for my shyness. Teaching has molded me into a more assertive and confident person. Teaching also gives me the opportunity to build meaningful relationships with my students. They teach me new slang but sadly, I still don't feel completely up to date. Teaching exposes me to different people from diverse cultural backgrounds and I've learned so much from them. Teaching gives me an audience for my corny jokes. Kidding aside though, teaching has given me memories that I will cherish for a lifetime.

Teaching has also given me what I had set out to find when I first started on this path. At the end of last year, a student who had been consistently coming to tutoring gave me a parting gift because she was graduating. She wrote me a very sweet message, thanking me for encouraging her and apologizing for crying to me when things became overwhelming. With the message came a plaque which read, "World's Greatest Teacher: Because 'Superhero' isn't an official job title." Now, when students ask why I became a teacher or if I regret my career choice, I'm almost tempted to point out that there aren't that many professions where I get the chance to be a superhero.

CHAPTER 2

FROM THE MIRROR TO THE SMARTBOARD

Angelina Ezratty

I had always wanted to be a hairdresser. From age 3 or 4, I'd imagined myself in front of a mirror with a blow dryer in my left hand and barrel brush in my right, wearing red lipstick, having long nails, and owning my own store. I have a natural talent for styling hair, as well as for making people feel better about themselves, which was really what drew me to the profession. So, how did I end up in front of a SmartBoard with a calculator?

I am the product of two *very* analytical parents who pride themselves on their mathematical abilities. That being said, it was only natural that growing up, I had a lot of "math confidence." It had always been my favorite subject. Although, neither of my parents went to college, and I had *no* intentions of being the first one in my family to attend, I knew that if I wanted to be a successful business owner, I needed to be fluent in the language of mathematics.

The Inspirational Untold Stories of Secondary Mathematics Teachers, pages 13–23.
Copyright © 2020 by Information Age Publishing

Being somewhat of a perfectionist, as well as the oldest of five, striving to have my schoolwork put on the refrigerator, nothing less than 95 was acceptable in my young eyes. My parents are also business owners and would emphasize that their job was work and our job was school, and after all, to be an excellent business owner, one must first be an outstanding worker. I took my job *very* seriously and this mindset stayed with me all through college. But, how did I end up in college, you may be wondering...

At the end of my junior year of high school, after receiving a 92 on the Algebra II Regents exam and learning I would be taking calculus, my mom had decided to propose a deal with me, "Ange, if you get a scholarship to college, you are going." Although this was more of a demand than a deal, it was one I could adhere to nonetheless because I understood the value of college. Even though I still wanted my salon, I figured that this "deal" would guide me to whatever path I was supposed to go on. Simple enough, right? NO!

When the college acceptance letters arrived in December 2012, and I received a scholarship to *every* school to which I applied, I felt like I was being ripped in half. I had my dad pushing me to follow my dreams, my mom wanting me to be cautious, and friends and family members giving me opinions that I didn't even ask for, to take into consideration. The magnitude of this decision mentally and physically weighed on me *every day* from December 20th to April 20th.

During the December holiday break, I visited various colleges on Long Island, NY, one of which impressed me because I liked what the program had to offer, and also that President Obama had spoken there earlier in the year. At the time it seemed like my best option. I was going to pursue my B.S. in exercise sciences through their comprehensive program, hopefully graduate in 3 years, and go to beauty school afterwards. My degree would be a "fallback." If I ever needed a different career, this degree was a gateway into any medical profession. Phew, what a relief!

Saturday morning, April 20th, 2013, I went outside to walk my dog and noticed there was a letter left in the mailbox addressed to me. "Angelina, you have been accepted to the Queens College Freshman Honors Program with an annual scholarship." At this point, I was so frustrated with forcing myself to go to college that I did not even want to look at the letter. I actually stuffed it in a drawer in my room and didn't even tell anyone about it. However, in the middle of what was at the time, Waldbaum's, in the produce section, I had my first moment of "divine intervention."

"Mom, I don't even want to go to college. I am going to Queens College. I will pursue my degree in business, be done in 3 years because of the credits I received in high school, go to beauty school, and then eventually open my salon. The school on Long Island is farther, I don't even want to work in the medical field, *and* although I got a scholarship there, $20,000 is a LOT of money to pay for uncertainty."

I was a pretty calculated 17-year-old. Mama could not disagree with my argument and I had *finally* made a decision: Queens College B.A. in Business and Finance for graduation in Spring 2016.

* * *

August 28ᵗʰ, 2013: The first day of school. Yeah, the campus was pretty, the people were nice, but *why* was I here? What was I going to learn from all of these people that would benefit me? Why didn't I go away to college? What else is out there? I miss all of my friends. Why am I doing this??

***

October 16ᵗʰ, 2013: "I'll join a sorority. I will meet people. I will make the most of this." It was sorority recruitment day and I am a *natural* at mingling. I encountered one girl, and we got to chatting. We had *a lot* in common, except that she was in an honors program to be a secondary mathematics teacher. I wanted to find out: "What is TIME 2000?"

For some reason, which I also attribute to "divine intervention," I felt *compelled* to meet the people who ran this honors program and become part of it. I had *no* intentions of becoming a teacher. I loved my high school teachers, especially my mathematics teachers, but I did *not* want to *become* them. So, I am sure you can imagine the frustration I felt—why was I wasting people's time? Why was I wasting my own time? *Why* did I feel like there was something else for me besides hairdressing?

From October 2013 to February 2017, I dealt with an internal conflict—"I want to own a hair salon, I don't want to be a teacher. I enjoy working at the hair salon every weekend. But, I also enjoy *all* of my undergraduate classes, and creating lessons, and helping students believe in themselves, and making students smile before leaving my classroom. What am I going to do?"

DEVELOPING A PORTFOLIO

After my freshman year of college, as a requirement of the TIME 2000 Program, we had to make a portfolio documenting the "highlights" of our year. What did we learn? How will that affect our teaching? How can we share these lessons we have learned with others?

I've always been a very self-reflective person, but I never did it in a structured manner, until I had to create my annual portfolios. After making the first portfolio and realizing what I had accomplished in that first year and what I had learned from it, I wanted to know how I could bring this type of reflection into my daily world. I started reading books and following Instagram accounts that had to do with essentially following your fears and landing right on the path where you belong.

After three years of learning how to follow my path, I still was no closer to making a decision than I was when I joined TIME 2000.

***

February 14th, 2017: At this point, I had student taught at a middle school and LOVED it, I had passed one of my requirements for teacher certification—Educative Teacher Performance Assessment (edTPA)—with mastery, and I started student-teaching high school geometry. This was my last semester, my last class; time was ticking.

"Good morning Angelina, I will be your cooperating teacher. This is what we are going to teach next week, so start preparing and send me what you have so we can make the edits." My cooperating teacher for high school geometry was very supportive. With that being said, it was not her, or even any of the people I worked with at the high school level. Rather it was the difference between going from an *intimate* middle school setting where I had three classes to focus on and knew the students well, to being in a large high school where I had five classes and did not have the chance to learn much about the students. In the middle school, students were placed in carefully selected groups, and instruction was data-driven and suited to *each* student's needs. In the high school, instruction was fast paced without any data-driven rationales and with no opportunities for remediation, which didn't sit well with me. I also just missed the familiarity of the middle school routines—we had a morning assembly, there were anchor charts in the room, we used lots of manipulatives and the students were receptive to my famous line, "Have a great day, good luck on everything!"

In this particularly large high school, the teachers and students worked on split schedules due to the school being so overcrowded. Because of this, one could feel the lack of *unity* within the school. Further, the middle schools in which I did all of my observations and prior student teaching were lottery schools in which the students had to apply, which indicated there was a lot of parental involvement. As a consequence, many of these students were *invested* in school, involved in afterschool activities and it was obvious that school was their main priority. The high school I was student teaching at was quite different. This was the first time I encountered students who were 20 years old and still in algebra, or witnessing dangerous fights in the hallways, or students coming to school high, or in the same clothes every day, or not coming at all, or talking back to teachers. Despite my own anxiety from being there, I formed connections with a few students, the ones all the other teachers clearly no longer wanted to invest in.

Additionally, over that January break, I had intentionally lost about 20 pounds and it seemed that once I entered the high school, my anxiety got the best of me and those 20 pounds and a few extra, were creeping back on.

After about a week of being at this large, old, urban high school, I started developing a horrible cough and was told I had asthma. As a child, I was on a nebulizer, but I had outgrown my asthma. Where was this coming from? It turns out, I was allergic to chalk—this was a sign if I ever saw one!

Viewing this situation with older, wiser eyes, I was a *little* dramatic, but at the time I really could not see past what was in front of me. My anxiety had completely taken over my existence those first two weeks of February. I would cry unexplainably and unexpectedly. I was not able to sleep at night. I was driving myself, my family, and I'm sure others, crazy! After about a week, my family and I came to an agreement that it would be best if I put my health and happiness first, in other words, stop student teaching, a vital, required step in obtaining a teaching license in the State of New York.

After two weeks of teaching high school geometry, I marched into my professor's office, sat right in front of her and said, "Thank you for all you have done, but I am not going to be a teacher. I am going to switch my major to strictly math, graduate, and figure it out from there. I am sorry for all of your time that I have wasted."

Looking back on this, I think that moment may have aged her a bit. Nonetheless, she responded in the way that only she can—taking in all of the information and carefully constructing her response. She called in one of my other professors and after talking some sense into me, we decided that I was going to finish the semester, as I was *so* close to finishing that it was truly illogical to change degrees at this point and if I didn't want to teach, that was something I could decide *after* graduation.

ANOTHER DEAL

During class that night, I was presented with an opportunity to participate in a "mock interview." A "mock interview" is an *awesome* experience that my professors organize where several school administrators are invited to interview a few of the future teachers in front of the *entire* class. We get feedback from the administrators, our classmates, and our professors, *and* we have some experience so that when we actually interview for jobs, we know what to expect. Being the over-achieving, over-prepared, person that I am, this was right up my alley.

"Any volunteers?"

My hand shot up. My two professors who had just convinced me to stay in student teaching, exchanged a knowing glance, and gleefully encouraged me, *thankfully!*

After class that night, I went home *so* angry. Why couldn't I follow through on my decision to leave? Why did I feel so *pulled* to be a teacher yet so sure that being a beautician was my path?

That week, I begrudgingly went to student teaching and did not feel any fonder about teaching than I did prior to my meeting with my professors. It wasn't that I didn't *like* teaching, or *like* students. It was more that I did not like feeling like I was watching *kids* slip through the cracks, becoming victims of circumstances that they couldn't control. Instead of thinking I could make a difference, which seemed like a godly task at the time, I thought running away and pretending it didn't exist was a better option.

My mom asked, "Ange, why even interview if you're not going to pursue this?" Honestly, I really couldn't answer her. This thought had also crossed my mind, but every time I went to email one of my professors about not going on the interview, I physically felt like I couldn't write the email; another "divine" moment. This interview was something I knew I *had* to do.

***

Tuesday February 28th, 2017
7:30 AM: A text message from my cooperating teacher, "G.m. [good morning] Angelina. My son is sick. You are teaching today. Good luck."

First I was nervous, then I was annoyed, and then I realized, "She trusts me." I opened every single blind, I let some fresh air in, I cleaned the boards, I lined up the desks, turned on some music and remembered why I was still even here.

8:05 AM: "Good morning students! Come in, start the do now…Whoever puts the homework on the board can get extra credit, but we only need 4 people, BUT if you can find another method *or* refine their work, you can also gain points… Let's start attendance."

"Ms. Ezratty, can you help me with #5?"

"Ladies, why are we late?" *Can I talk to you outside?*

Now I remember why I wanted to teach.

That afternoon, I had my interview. Everyone else who volunteered was *freaking out*; each wondering which administrator would interview them. I, on the other hand, was not worried. I was going to be myself and do my best and just *learn* from the experience. That was my only goal.

4:30 PM: As I entered the familiar doors of Powdermaker Hall, I saw a *giant* bald man who reminded me of my father. He was dressed to the nines, and very engaged in a conversation with my professors. I felt some buzzing of excitement that I still can't explain.

4:35 PM: As I was sitting in class, waiting for my interview, I hoped that this man, who I later learned was the assistant principal from a local middle school, was the person with whom I would have my mock interview.

4:50 PM: The giant bald man himself, Mr. G, comes walking in and I hear, "Angelina, it's your turn." I nearly *flew* out of my seat.

***

I'd be lying if I tried to recall the words that were exchanged during our interview, but I distinctly remember the feeling. *Every* question that was asked, I felt like the answer was already picked for me and I was just a vessel to deliver the responses. There is one moment that I do remember:

"Why our school?"

"Honestly?"

A nod.

As Aunt Deb would say, "It came out like a burp."

"When I was researching your school online, I was sitting at my computer going, 'This is *so* cool...oh my gosh, they have that?? WOAH!' and my sister kept saying, 'Ange, what?...what?' to the point she got up and I finally noticed she was talking to me and I said, 'This school that I could be interviewing for is SO cool. They *literally* have a phone number that students *and* parents can call to get homework help from in *any* language!' and my sister responded, 'Only *you* would find that cool.'"

Another nod, like "keep going."

"Just from my research on your school, I could tell how much care and support there is, not just for students, but for their families as well, and that type of support is aligned with who I am and what I believe in. I also saw that you have an overwhelmingly Spanish population and I am teaching myself Spanish, so maybe I could communicate with some of those parents one day."

At this point, Mr. G stopped the interview. He turned to the class and said, "I am being *completely* unorthodox now by stopping the interview, but I must tell all of you, *that's* how you go on an interview."

So, what happened?

Mr. G and I kept in very close contact and he is now my boss. I essentially got hired that night and there was no turning back.

***

March 24ᵗʰ, 2017: When I went for my official interview, I walked into the school and was *so* impressed by the entrance.

Welcoming.

As I entered the office, it was clear that *I* was known. It had been less than a month since my mock interview with Mr. G, but it seemed that this day was much anticipated, on both ends. *What* a feeling.

After my demo-lesson, which was done in the room that is now *my* classroom (I HAVE MY OWN CLASSROOM AS A FIRST YEAR TEACHER!), Mr. G. and I had a discussion about the lesson, more of a *reflection.* I was very prepared for this, as my professors started getting us used to this type of one-on-one dialogue since freshman year. He then said, "Let's go meet the principal. Do you remember what you said at our interview?"

I said a lot of things.

Almost as if reading my mind, Mr. G said, "*Why* do you want to be here?"

Oh yeah!

"Well, I have new and better answers since then."

"I'm listening."

"This school is *so* welcoming! You have *chairs* in the lobby! You have double staircases; you have student work hanging up! You have student-made artwork lining the walls! You have *plants* in the office! This is a place that *cares*."

We were ready to enter the principal's office.

***

I *immediately* emailed my professors:

On Fri, Mar 24, 2017 at 1:15 PM, Angelina Ezratty wrote:

I AM --------------------THE NEWEST EMPLOYEE FOR FALL 2017!!!!

Thank you all for everything. I have so much gratitude for each of you and am so excited for this.

—*Angelina Ezratty*

***

I finished student teaching for college credit the last week of May and rather than finishing out the school year at the high school, I asked Mr. G. if I could start going to *our* school. I would not be getting compensated for this, but I also was not getting compensated for being at the high school. I felt that it was better off in the long run to start meeting my colleagues, learning the daily routines of the school, and even just knowing where the bathroom was *then*, rather than when I had 120 students in front of me.

During this time, I was lucky enough to be under the wing of a fellow TIME 2000 alumna. "K" openly invited me into her classroom, even trusting me to co-teach with her. We became good friends and "K" continued to help me throughout the summer and we still go to brunch once a month and work closely together.

Although my month of June did teach me where the bathroom was, there was so much I couldn't anticipate. How does running a homeroom work? When and how do I receive my laptop? How do I go about making copies? When will I stop getting new students each day? What are my responsibilities? It was *so* much more than lesson planning.

***

In the early days, I felt like I needed to work *all the time*. If I wasn't working, I felt guilty. I was teaching remedial math at the time, so I did not have a curriculum. I had to lead instruction based on the feedback of my students' core mathematics teachers, as well as the work they were producing. It took a few months to start balancing work, graduate school, and life.

I had one class of English language learners (ELLs) who did not speak *any* English when we first met, one class of 13 self-contained students, one inclusion class *without* a co-teacher, one "behavioral" class, and one on-grade class. There were a lot of student needs that I wasn't sure I could meet.

As we got to know each other, we fell into a flow. Having a remedial section of classes was actually a blessing in disguise. I was able to experiment with new technologies, figure out what works for me and what doesn't, as well as be involved in other areas of the school, like going on trips. I have gone on almost every field trip this year, even leading my own to the New York Stock Exchange after playing the Stock Market Game with my on-grade class.

The school grew as *I* grew. During my first observation, which was an informal, which means I was not expecting it, I had written on the board, "F.G.F. What do you love about yourself?" F.G.F. stands for "Feel Good Friday." As mentioned previously, I enjoy self-reflection and want my students to learn to love themselves. I started doing F.G.F. my first week of working at the school. The first F.G.F. was, "What was the *best* part of your week?" I had *no* idea what this little closing question would do.

Looking back, it really is not that surprising that I wanted to teach my students how to self-reflect. Given my anxiety and battles with uncertainty, I needed some form of outlet and writing had always been there for me. When I couldn't understand what was going on or what I was feeling, I would just grab a notebook and write whatever *flowed*. It was almost like when my pen hit the paper, time stopped. I was able to stop worrying. As I got older, I found more structured ways to do this, but I didn't realize the true *power* of reflection through writing until I did my first portfolio for TIME 2000.

I remember sitting at my dining room table with *all* of my work from the past semester and thinking, "Where do I even begin?" However, just beginning was enough of a first step. As I started sifting through the materials, the memories came back, and the words poured out. I found the whole process *enjoyable*. I started to look forward to this yearly reflection, even telling my friends outside of the program that they should try it.

My enthusiasm for this project and the care I put into each portfolio allowed me the opportunity to actually host a portfolio fair for younger members of the program. This was the first time I shared my portfolio with anyone other than the faculty and it was truly a vulnerable, but valuable, position. I learned that by being my authentic self and sharing my triumphs and tribulations actually made me more *human* and as a result, allowed me to help others on their own journey. Many students in the program reached out to me and I was happy to help.

By February 2018, Mr. G invited me to a Core Inquiry meeting, which is basically the school improvement team. I was 22 years old at the time, had not even been working there for 6 months, and was asked to present about F.G.F., without any notice, to the *entire* team, including the principal. By the end of the meeting, my principal looked at me and said, "You're on the team."

As I keep referring to it, one of my core beliefs is continuous self-reflection and improvement, as well as recognizing that failure is an integral part of success—that failure is really just a synonym for *learning*. Throughout the year, I would peruse the internet and save materials I liked, as I knew one day I could use them, even if I didn't know what I could use them for *yet*. At the Core Inquiry meeting, I was asked to be on the Growth Mindset team. Well, I just so happened to have saved *100* Growth Mindset posters earlier in the year. I thought 100 was *way* too many to print, but I didn't want to delete the file. After the meeting, I emailed Mr. G. the posters. By the following meeting, the posters were printed, laminated, and hanging in the hallways. *We were growing together.*

***

Tuesday June 5th, 2018: "Good afternoon staff. Please check your mailboxes before the close of business today to retrieve your programs for the 2018–19 school year."

Deep breaths. What grade will I teach? Will I have seventh-grade remedial again, or eighth-grade algebra? I know I did well teaching eighth-grade algebra in Saturday school, but was my discipline style suitable for eighth grade? At this point, I've learned to look my anxiety in the face and say, "Thank you for your opinion, but I can handle and will succeed at whatever comes my way, but thank you for your concern." With that little mantra in mind, I grabbed my program and was overcome with joy—seventh-grade *core* mathematics. I was *trusted* to be the initial deliverer of instruction.

Teaching a program of remedial classes my first year, followed by teaching core classes of the same grade the following year is actually a *blessing*. During my first year, I was able to learn the misconceptions of my students regarding the seventh-grade content, experiment with the rituals, routines, and classroom management strategies I wanted to use and could now implement all of this knowledge for my new program. Although I will be armed with one year of experience, I have new challenges to learn from—this will be the first time I have a co-teacher every day, as I had a co-teacher once a week for my ELL class. I will also have to adhere to a time-restriction and prepare my students for the state exam, as well as their midterm and final. I know this will be challenging, but I know I would not be asked to take on something I couldn't handle.

In the "math world," things tend to come full circle. Around the first day of school, I had to take off to attend a funeral. It so happens that the second to last day of school, I also had to attend a funeral. This is another "untold story": balancing life and emotions along with work. Many people advise me not to tell my students about my personal life, but I think it's important for them to learn how to handle these situations. On the last Friday of the year, I said, "Ladies and Gentlemen, unfortunately I have to attend a funeral on Monday so we will be doing our last day of school activity today." *I was met with nothing but compassion.* Our last day of school activity was for each member of the class, including me, to write our names in the center of a piece of paper and pass it to the other members of the class either writing a favorite memory, a well wish, or something similar. When I received my paper, I was moved to tears:

Thank you for being there for me.

I came to grow as a successful student because of you. Thank you.

Thanks for being the most supportive teacher I had this year.

Thank you for getting me interested in math.

These comments actually motivated me to compile a portfolio this year—something I grew to *love* doing as an undergraduate; *full circle*.

Who I am and where I am today would not be possible had I not been lured by the attraction of being in a special program for people who had the same interests as I did. Between the introduction to self-reflection through portfolios, the one-on-one support and relationships of my professors, and the opportunity to interview with Mr. G., I have ended up right where I belong. As for hair, I still work at the hair salon if they really need me and I have the time. I do braids for dance recitals, and I keep in touch with old clients. My admiration for the hair profession has not changed, but the urgency of feeling like I need to be there has subsided. I am right where I belong, in front of a SmartBoard with a calculator.

CHAPTER 3

MY UNEXPECTED HAPPINESS

Daniel De Sousa

The period is about to start, and I greet each of my students as they come in. "Jessica how are you today!? Wow, Myrone, did you get new sneakers? I love them! Ladies and gents, as you come in, please pick up your graphic organizer for the day. Thank you!" You would think that setting this pleasant classroom environment would ensure proper student behavior. Not true! No matter how nice or *cool* I was to my students, they would enter the room and have off-task conversations with each other. My plan backfired. I gave them too much freedom. Fewer than half of my students actively participated in any of my lessons. What should I do?

Flashback to the field observations I had during my teacher preparation program: The field work was unique in that our whole college class met with one teacher who spoke with us before and after we observed her lesson. I recalled one instance when we discussed behaviorist strategies. The teacher shared her many approaches (e.g., reward systems, routines, etc.). She also modeled these approaches and I remembered how she would stand at the door and urge the stu-

The Inspirational Untold Stories of Secondary Mathematics Teachers, pages 25–30.

dents to come in, take their seats, write down the aim, the Do Now, and begin working. Generally, there weren't many students off task, and that was partially due to the reward structure she had created for them. This is what I realized was missing in my class, an incentive for my students to behave. During the observations, it was obvious that the middle school students love stamps and stickers. Who would have guessed that my high school students would love them, too? Much to my surprise, when I implemented this strategy, all the students got on board, behaving and staying on task.

***

"Daniel De Sousa," my diploma said. I graduated with a degree in "Mathematics and Secondary Education." My dream came true. I immediately received a job as a high school geometry teacher in Queens, New York. I'll be honest, although I felt fully prepared and full of energy each day, I was intimidated at first. I was always worried about not doing the *best* I could. It wasn't easy getting to where I am today.

As a child, I hated learning mathematics. Thinking back to when I was in middle school, any time someone asked me whether I liked mathematics, I always responded with, "It is so boring and not interesting." I really struggled with mathematics up until high school and even then it took such a long time for me to appreciate it. But, when I was asked what I wanted to be when I grow up, I always answered that I wanted to become a teacher. From an early age, I always took the time to help my friends and classmates. I remember I loved to sit with my friends and pretend to be the teacher. Some called me a teacher's pet, but I didn't care. I took that as a compliment because I wanted to be the teacher. I remember helping my friends with our English homework, and history projects. I would even go home and *play school* by teaching my mom what I learned. However, there was one subject I stayed away from in my early school years: Mathematics.

My love of mathematics blossomed in high school. For once, I had a mathematics teacher who didn't just tell us how to do mathematics, she taught us the reasoning behind the procedures we were learning. This was the first time it all made sense to me, which is when I knew that I needed to be a mathematics teacher with the mission of changing students' perceptions that mathematics is boring and useless.

Life was hard. It was always hard. I grew up in a one-bedroom apartment with five people living in it including myself. I dealt with many difficult family issues throughout my life which for the most part were out of my control. My father worked as a plumber, my mother worked as an aid in a private school, and my two older brothers, at the time, were alcoholics and drug abusers. My brothers never worked hard. They caused me so much pain that even today, I still vividly remember all of the disturbing incidences that occurred. My father was never home so I always had to look after my brothers; a 16-year-old taking care of his adult brothers. Throughout my high school life, I was depressed. I'd go to school and then come home to my brothers who were either strung out or drunk. There were times when I had to stay up late or not even sleep at all because I had to make sure they didn't kill themselves. I am not sure whether my brothers realize it, but many years later it was because of me that they got help.

Needless to say, a lot was expected of me at home and at school. I had to make up for the shortcomings of my brothers. I had to be successful, but I knew that more importantly, I wanted to be happy; and what would make me happy was being a mathematics teacher.

When I finished high school, I applied to the TIME 2000 program and I met the director at my high school. I was eventually sent a letter informing me that I was accepted into the program. I could not have been happier. I started telling my friends and family about my good news. See, I am one of the first in my entire family to be attending a four-year college and the first to attend college in my immediate family. This was so important to me. This was what was expected of me and I made it. Little did I know then, that my difficult home life along with my preservice field work would serve me well as a mathematics teacher enabling me to understand and support my students, Peter and Isabel.

PETER: ATTITUDE AND ACADEMIC CHALLENGES

Now, getting back to my geometry classes, I was just starting to get into the swing of things. I had my routine of having my students coming in nicely and my students would be on task. I had hand signals set into place to initiate when I needed attention. Everything felt like a well-oiled machine.

At this point in the lesson, students were working in groups while I went around and listened in on their conversations in case I needed to chime in. I remember my assistant principal summoning me to meet her in the hallway for a "second." Luckily, my co-teacher was in the room, so I left to see what she wanted. I went into the hallway and I saw a student almost as tall as I am standing behind my assistant principal.

"Hello Mr. De Sousa! This is your new student, Peter," she said in a cheery voice.

"Wow! Peter it is so nice to meet you. Please come in and we can work together to catch you up with what we have been doing in my class. Don't worry," I said with enthusiasm and a sincere smile.

"Nah. Nope, I'm not going in there," Peter stated as he walked away from us.

"Don't worry, I will bring him back," said the assistant principal.

About 5 minutes passed when Peter came into my room looking around. I walked over to him and I pointed to where he could sit, but I did it with a lot of enthusiasm and excitement hoping that I might be able to "break the ice" with him. He stopped dead in his tracks and said stone cold, "Nah, don't do that." I felt intimidated because I had no knowledge of his background and I did not know how I could reach him. As my co-teacher took over, I gave Peter a syllabus and other starting materials. I tried sitting with him to make him feel comfortable and included in the class, but he wanted no part of me. I felt that I needed to give him space, so I explained what was going on in class and I assured him that if he needed me to let me know. I walked away and under his breath he had a few inappropriate things to say to me. In that split second, I needed to make a decision. Do I turn around and say something, or do I leave it and walk away?

That was one of the best things about our field observations. We would usually go into an observation with a set idea/strategy that we were looking for. But the beautiful gems that you can take out of an observation are sometimes the things that you weren't necessarily looking for. I remember a situation when the teacher we were observing dealt with a small argument between two students. She stopped what she was doing to address the issue, but the issue didn't get any better. Later at the post-teaching discussion, this event was brought up. The teacher remarked that she shouldn't have addressed the students because they weren't ready to resolve the problem. At the time, I didn't understand why a teacher would ignore unwanted behaviors.

Recalling the observation, I walked away from Peter. Now knowing what I found out, it was the best thing I could have done. All teachers have had or will have a "Peter" in their classes. He was a student who would get up and leave my classroom, bad mouth other students, not listen to authority, and sometimes come in smelling like weed. It was incredibly hard for me. This student was quickly trailing behind, farther each day. I wanted to help him, so I asked my colleagues about him. Most of them just said, "That's Peter. At least he is better than last year. You are lucky." While I knew that they meant this in a way to comfort me, that wasn't the answer I was looking for. It took a very long time to figure out how to help this student. I wasn't going to give up, just like I wasn't going to give up on my brothers when I was a child. Luckily, I was able to work with my principal who was beyond helpful with figuring out ways to reach Peter.

It took months to change his negative behaviors. We focused on Peter coming to class and taking notes. To some this may seem as though he was taking "baby steps," but for Peter they were giant steps. You see, Peter was not used to a classroom setting like this. It was hard for him to feel secure and confident in a setting with thirty other students. I found out that before high school, Peter used to be in a self-contained setting in his previous school. There were fewer students in a class and Peter was able to have more attention. However, being in a classroom filled with about 34 students, most of his needs were not being met. My colleagues who noticed Peter's progress, were impressed.

In my school, we have a system where we write anecdotes about students to describe their behaviors. These anecdotes contain positive, negative, or neutral comments used for record keeping and are shared with other faculty members and administrators. Students get benefits for documented proper behavior while students who would lose points would have to *restore* themselves in a restorative justice program at our school. It is similar to what I observed in fieldwork when students were selected to be the "Student of the Month." I will never forget the time when I wrote up a very favorable anecdote for Peter describing his good behaviors in my class. I was told by a colleague that one day when Peter went to the school store to purchase some food, he was denied because he had too many demerits. Peter was very angry and said that all his teachers write only bad things about him. My colleague told him that not all teachers do. She proceeded to read a few of my anecdotes to him and said, "You know Mr. De Sousa wrote that about you, right?" She explained to me how Peter smiled and said, "Mr. De Sousa wrote that about me?" From then on Peter and I had a much better relationship and it's because I took the time to document his good behavior and it came to his attention. Sometimes the most difficult students end up bringing teachers the most satisfaction.

ISABEL: EMOTIONAL CHALLENGES

My first few months of teaching, although challenging, were the most enjoyable. I had my own classroom and I was a mathematics teacher. That is all I wanted because I knew happiness followed. Classroom management has always been a challenge for me, and I suspect it always will be. Some may say that looking into your students' emotions and caring for your students' mental and physical health has nothing to do with teaching mathematics and managing a classroom. But, I learned it has everything to do with reaching all students so that they can understand mathematics.

It was the mid-school year and I remember finishing a lesson on quadrilaterals. During the next period, which was my lunch break a student, Isabel, came into my room. She seemed in distress and eager to talk to someone. As always, I tell my students that I am there for them, but this was the first time a student actually came to me about something.

"Mr. De Sousa, how am I doing in my class?"

"How do you think you are doing?" I responded curiously.

"Well, I don't think I am doing well anymore. I mean, I was doing well in the beginning but now I feel lost and I am going to fail your class, your test, the Regents exam, LIFE!"

I was taken aback a bit. This student was literally crying out to me. I remember when I was doing my fieldwork, the teacher noticed a student who was not being his normal self. I recall her talking to him quietly. It was non-threatening and seemed like she was just having a normal conversation. I noticed that she gave him a job to do in class which, using what I know now, allowed him to disconnect with what was bothering him at that time. To this day, I don't know what she said, but what I do know was that she helped him. She used a mixture of positive reinforcement and giving him responsibilities that made him feel important. This memory came back to me as I dealt with Isabel.

"Isabel, you are one of the most active students in my class. You are doing so well and you have been showing me that you care a lot about your work in my class. Why on earth would you think so negatively about the progress you have been making?"

"My parents told me I am a failure. They are ashamed of how I have been doing, and I just—"

"Failure is not a word that describes you," I interjected. "Sometimes, our families can give us expectations whether they are obtainable or not. Sometimes an expectation is too high and is not realistic. What do *you* think of yourself?"

"My parents believe that I slack—"

"No, Isabel. What do *you think about yourself?*" I responded calmly.

"I try my hardest in your class. I may not be the smartest but that is why I ask for help from Mike. I know he is the smartest kid, and he helps me a lot. I don't know what to do. My parents don't think of me highly and I just want to give up. Why should I keep trying if it doesn't make a difference?"

"I always think of you highly. I have seen so much progress in your abilities this year. You have come such a long way. Let me ask you, what do you want to be when you graduate?"

"I kind of want to go into the science field."

"So, that is your goal and that is what you are going to strive for. Listen, we are going to work together more this year and I am going to help you prepare more for what you want to do when you leave high school. I'll talk to Ms. B. She has had a lot of jobs in the science field before she became a teacher."

"You would do that for me?"

"Of course! I am going to ask you to do a few things for me during class that I think you will benefit from." I thought back to my experience from my observations and decided to get her to do more things in class so that she would feel valued and distracted from what was going on outside of school. "You know how Mike helps you when you are stuck? Well I want you to do the same, but I want you to help out Michelle in class when she is struggling. I am going to change seats soon, and I want you to be there for her to support her. I appreciate that you came to see me!

Needless to say, I never got to eat lunch. However, this was more important. This student wound up doing well throughout the remainder of the year. She always came for help and started helping others in the class. She passed the Regents exam that she was so terrified of and is now moving on.

Among other students, the Peters and the Isabels come with their own sets of challenges. I look forward to supporting them and helping them overcome their problems, and rejoicing with them in their successes.

CHAPTER 4

A JOURNEY IN DEFINING MY INNER TEACHER

Irina Kimyagarov

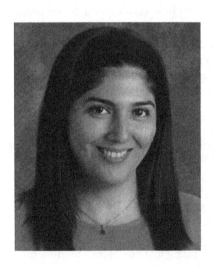

My knees were knocking, and my voice was shaky. I felt my face flush with embarrassment, and my eyes were swelling up ready to burst into tears, waiting for the moment that I could sit down and hide from everyone. This was an all too familiar feeling; one I've grown accustomed to every time I have had to take my place in front of my peers to present a class assignment. Everyone was staring, waiting for me to say something interesting. I froze in place, staring back, imagining their judgement of me, my intelligence, abilities, and skills. This is exactly why I never spoke up in school or volunteered for any class presentations. That was high school. What did I possibly think was going to change by the time I was back in high school, as a teacher, once again standing in front of a group of teenagers, trying to present what I know?

What was I thinking going into teaching? Why would I subject myself to this same feeling of anxiety day after day?

As I write this, I have completed my fifteenth year at a suburban junior-senior high school in Long Island, New York. I've spent six of those years as a mathematics teacher, having taught the full gamut of classes from remedial to advanced placement (AP), and the rest as the chairperson of the mathematics department. We serve a diverse minority population with nearly half qualifying for free or reduced-priced lunch. Our students experience successes that exceed the expectations of the outside world, and I've been lucky enough to work in that same school since I started my career. How did I get to a point where I not only teach teenagers who look at me with that same expectation of interesting content, but also supervise and mentor seventeen mathematics teachers, who look to me for guidance on a daily basis?

I can tell you the exact day and moment when I felt the change. It was the first annual TIME 2000 Conference, *Celebrating Mathematics Teaching,* in the fall of my senior year in college, and I was at the podium about to lead the student forum in front of nearly 100 high school students, their teachers, other students in the program, and our professors. That was the moment I felt that change—the moment I felt that I can, without a doubt, stand in front of a group of students, or even teachers, and lead them in engaging learning. At that moment I stopped thinking about what my audience was thinking of me and began thinking about how my words could make a difference for them. The journey leading up to that point, and the equally critical journey that followed was one of defining my inner teacher. Much of what happened in preparation for that conference, and the day of the event has had an impact on me and my career as a mathematics teacher. My high school self would not even recognize me today. I sometimes still look at myself in disbelief of how far I have come.

I spent my formative years shying away from the spotlight. I spent so much time trying not to be noticed that I didn't engage in experiences that would help me mold my identity. I had a very small circle of friends. My focus was on academics and I preferred to work on my own. Consequently, the search for "self" was delayed and ended up spilling over into my early college years.

Despite the many professionally defining opportunities that I had as a pre-service teacher—fieldwork, conferences, and related coursework—I didn't have the headspace to define a professional self until I defined "me." It should be of no surprise then that it was a single moment during that conference in my senior year that everything finally clicked; I figured out who I was, and so I was ready to embark on the journey of defining my inner teacher.

So, who did I turn out to be? The most surprising of discoveries was that I thrived in leadership roles. How ironic! The very thing I avoided during all of my years in school was the one thing I most enjoyed about the work that I did. I enjoyed working with others, often taking charge in group tasks, facilitating problem-solving discussions. I took a college assistant job for my undergraduate program, TIME 2000, working in its office and assisting in its day-to-day administration. I savored opportunities to work on developing various programs—working out the logistics of activities, and events. I held a position of club president, and later ran for student government at the college. I found my passion. I found what I was good at. I found "me."

I remember my years as the college assistant to the TIME 2000 Program with great reverence. This experience has had the most profound impact on me as a person, and as a professional. It helped me mold my identity, build confidence, and aspirations.

Running the program office was uncharted territory for the three of us, all hired around the same time: a secretary, an administrative assistant, and me. I was least experienced out of the bunch but was determined to be useful. My first task, as I remember it, was to digitize student files so that data recording student progress could be retrieved easily. That was well beyond my skill set, and long before cloud sharing. I'm sure someone figured it out by now because I definitely could not then. So, my first task, as I remember it, was an epic fail!

Every experience in those three years was a first for me. Most were met with success. I learned a lot about myself, about working with and for others. Most importantly, I learned how to keep a professional line. At times I was privy to information that could not be shared with others, especially my program peers. I had access to student applications with personal statements and high school grades, along with other sensitive information that was not meant to leave the office. I learned the importance of confidentiality and of trust. I was trusted enough to be part of the behind-the-scenes work of a specialty program at Queens College. I wasn't going to compromise that in any way.

The program's administrative assistant was someone I came to consider as a mentor, friend, and pseudo mom. We worked well from the very beginning. There was no project too large or too hard that we couldn't tackle together. For example, one day the program director, came in, sharing her dream of an annual conference for high school students. The idea was for high school students to experience inspirational mathematics lessons that might motivate them to consider becoming mathematics teachers themselves. We set out to make that dream come true to the best of our abilities.

I remember sitting down with the program administrator and charting out what we would need for the conference. A venue--Some place on campus large enough to hold an audience of 100 or 150 students and their teachers. A keynote speaker—Someone to really make a memorable impression on the students. Great teachers from the field—To share their favorite lessons. What was going to make this conference great were the details (e.g., the lesson descriptions, the responses, job descriptions for TIME 2000 helpers, etc.) We didn't want to overlook any of them.

We brainstormed every aspect of the conference. What time did we want students and their teachers to come? How did we want them to check in? Do we serve breakfast? Does our venue provide space for that? How would the students know which sessions to go to? Do we have them choose or pre-assign? Do we want them to travel with their school mates, or should we separate them? How would they find the various locations on campus? What about the presenters—how do we check them in? What can we give them as "Thank You" gifts for sharing their lesson ideas at the conference? We want to inspire high school students to consider the math-teaching profession, so how do we educate them about the option of attending the TIME 2000 Program? Whom do we put on the student panel to speak about the program? Whom should we select to moderate this forum? These, and many other questions were asked and resolved as we planned through the spring and summer months.

I remember learning mail merge for the first time as we personalized students' agendas for that day. The nuances of sorting and filtering in Excel were also discovered. Mind you, this was really well beyond what I had ever used computers for. I was learning how to use the various Microsoft Office programs for administrative work, to facilitate large scale data manipulation.

I also remember what I then considered to be the most innovative idea on my part. As with everything we did in TIME 2000, there had to be a reflective piece. We wanted feedback on the conference: things that went well, sessions that were well received. Including a survey in the students' and teachers' welcome package was easy, but how do we get them to fill out the forms and submit them? This was crucial. We wanted this to be an annual conference and it needed to change and improve as it grew. I got it! That survey was going to be their lunch ticket. Literally. We had program students standing with boxes at the lunch lines, asking students for their survey as the ticket to get lunch. Who wouldn't fill out a survey to get lunch? The return rate on that feedback was 100%!

The program administrator and I put in months of hard work into this conference. With every day I found myself more confident in my contributions to its development, more proud of being a part of this great project. I wanted nothing more than to see the fruits of our labor. That morning I woke up extra early, dressed in one of my favorite outfits, did my hair and makeup, grabbed a large cup of coffee and made my way to the college. I was running on pure adrenaline—so excited that this day had finally come. We had invested so much into this; I had invested so much of myself into this conference. So many elements of its plan and execution had traces of me that I felt such power and control on that day.

As I stood at that podium in the fall of my senior year, ready to address the audience, I saw one of my professors gesture to me from the front row. The college president arrived and wanted to address the conference attendees before we began the forum. I took the liberty to introduce him myself, later learning that my professor had intended to do so. Oops! I guess I took my leadership a little too far. I hoped the president didn't get offended...

So here I was, finally starting to think of myself as a teacher. Through my student-teaching experience, I learned that I didn't have the nurturing empathy required to teach middle school students. They were cute children, but I just couldn't connect with them. When I started as a fulltime teacher the fall after graduating, I taught high school juniors and seniors. Over the subsequent six years, my program consisted of a wide range of courses, and student skill sets. I had a great rapport with my colleagues, a supportive chairperson, and the respect of my peers. Despite all, I struggled to find gratification in my work. I thought that perhaps teaching more rigorous courses would bring me that satisfaction, and so I asked to teach AP courses. As much as I enjoyed the content and working with the students, I still found myself dissatisfied with myself as a teacher. Having completed my master's degree, I sought a post graduate degree in school leadership. I kept on trying new things, doing more, just to fill the void I was feeling. I couldn't figure out what it was that I needed finally to feel like I found my place as a teacher.

Completely unexpectedly, just as I was completing my school leadership degree, my chairperson announced his resignation from his position. He moved on to a higher position in a different district, creating a vacancy in our department. I thought about it—This was no coincidence. This was a door opening just for me. What have I got to lose? This was once again another advancement for me to try out in an effort to find my place. I applied for the job, and within a couple of days of the interview, accepted the job as chairperson of mathematics.

Needless to say, the first year as chairperson was intense—as intense as my first year of teaching. My first year I had three brand new tenure-track teachers. Their success was going to be a reflection of my supervision—much the same as our students' success is a reflection of our teaching. I worked hard and took great pride in the work that I did with them. I approached each observation with a plan, much like a lesson plan. Each meeting with them had goals and objectives. I reviewed their lesson plans and provided timely and appropriate feedback—similar to what we learned to do for our students. I looked forward to seeing them progress over the course of their first year, setting goals with them for their subsequent years. Writing their tenure letters were moments of great reflection, and satisfaction. I finally figured it out; I found my inner teacher. I am a teacher of teachers.

Since becoming chairperson of the mathematics department in my school, I've enjoyed working with my teachers. I'm still a mathematics teacher, and my class consists of seventeen "students." As with any group of students, skills are always varied. I have the excellent teachers who need to continue being supported and challenged to be even greater. I have the teachers who need more of my attention and guidance. I have those who needed to be coached to proficiency, and yet others who needed an out-of-the box approach.

I've continued my work as a mentor and a coach of mathematics teachers outside of my department as well. I've taken on a position of a field supervisor for the Initial Clinical Experience for preservice teachers, guiding them through their first steps as student teachers. I've worked with my district coordinator to develop in-service courses for district mathematics teachers, as well as other professional development programs. I found my passion and my place.

My inner teacher is a teacher of teachers. All of those leadership opportunities from my undergraduate studies prepared me just for this. I didn't realize it then; I didn't have the headspace to understand it as an undergraduate. Thinking back, it was that Friday afternoon in the fall of 2002, at the first annual TIME 2000 Conference, *Celebrating Mathematics Teaching*, standing at the podium that everything fell into place. It just took me many years to find it.

CHAPTER 5

DARING TO LEAD

A Story of Early Leadership Development

Mara P. Markinson

I kicked the radiator in frustration; it was the dead of winter and freezing cold in the Buffalo hotel room. Exhausted from a long day of teaching followed by my flight, I searched for another blanket and then opened my laptop to anxiously key through the slides of my PowerPoint for what seemed like the millionth time. The next morning would be my first solo presentation at a professional mathematics teaching conference. Letting my insecurities get the best of me, I spent the next few hours unable to sleep. My mind racing, I pondered the same questions again and again: *What if nobody comes to my presentation? What if I make a mistake? What if someone asks me a math question I can't answer? Who do I think I am? Why won't the heat work?!*

The Inspirational Untold Stories of Secondary Mathematics Teachers, pages 37–44.
Copyright © 2020 by Information Age Publishing
All rights of reproduction in any form reserved.

Before I knew it, the sun began to peek through the hotel curtains. It was time. I took one deep breath after the next as I went through the motions of getting ready and locating the room where I would give my presentation. As I passed the large "Annual AMTNYS Conference" banner in the hotel lobby, I still couldn't believe that I was actually there to give a presentation for the Association of Mathematics Teachers of New York State (AMTNYS). I found the room where my presentation would be held and nervously stepped inside. Its size was intimidating; there were endless rows of chairs. With the questions from the night before racing through my mind, it took every bit of courage I could muster up to approach the podium and set up for my presentation.

As my hands clammed up, I faced my back to the rows of chairs and tried to calm down. I connected my laptop to the projector and the title of my presentation, "Conquering the Core with iPad Apps," filled the gigantic screen in front of me. I took the index cards highlighting my key points out of my pocket and scanned through them yet again. Hearing the hushed voices of people entering the room behind me, I knew I would have to turn around and face my growing audience. To my surprise, I turned around and saw that the room was already almost full. An audience member quickly approached me and said, "I am SO excited for this presentation! I've been trying to implement this in my classroom for a long time. I feel so fortunate to have someone to learn from." Not wanting my anxiety to show, I smiled and shook her hand.

I thought of my professor's reassuring words from the text message she sent me the night before. *You will be amazing! I wish I could be there.* I took a deep breath and began. I welcomed the audience and thanked them for coming. The room grew quiet as the conference attendees eagerly looked to me to guide them through an hour of professional learning. *You can do this,* I told myself. *It's too late to back out now.*

My anxiousness slowly but surely faded into the background as I settled into my presentation. I was fueled by the positive energy in the room—People were *genuinely* excited to hear what I had to say. As I engaged the audience in learning how to teach secondary mathematics lessons using iPad apps, participants were eager to contribute to the brewing discussion in the room. Soon, the room was full of life. Teachers were split into groups and were brainstorming ideas with each other, asking my opinion about how to change their traditional lessons into lessons on the iPad, and volunteering to participate in front of others to try out some of the activities I had planned.

Fifteen or so minutes into my presentation, a man quietly entered the room and took a seat towards the back, clearly trying not to interrupt the flow of what was going on. I approached him to prompt him to join a group, and abruptly stopped in my tracks. This was not just any stranger. Sitting there with a beaming smile on his face and the same twinkle in his eye that I could never forget, was my older brother's 6th-grade teacher. "Mara Markinson," he said, "You have no idea how much joy this brings me." Speechless and surprised, I forgot about everything else going on at the moment and stuttered, "What are you doing here? How did you know I was presenting? How? What?" He said, "Why don't you finish your presentation, and we'll talk after," as he winked and gave me a reassuring pat on the shoulder.

I caught my breath and stepped back into my presentation, so excited to talk with him afterwards. The rest of the hour flew by, and participants lined up at the end to thank me, to ask for my email address, and to shake my hand. Several audience members asked whether I would come to their schools to lead professional development for their mathematics departments. I couldn't believe what I was hearing. I was only 22 years old, and in my first year of teaching!

When the room cleared out, my brother's former teacher and I walked to the hotel restaurant where we sat down for lunch. "So," he said, "How on earth did you get here?" Knowing full well the immense struggles I had with mathematics as a child, and not having seen me for many years, it had to be equal parts gratifying and confusing to him to see me in this position.

As he already knew, I spent my childhood plagued with math anxiety— I never would have believed that mathematics would be in my future. I worked hard and memorized procedures to earn high marks, but I never understood anything that I was doing. What was worse was that none of my teachers *believed* that I didn't understand the concepts. I felt that my repetitive questions were annoying, and eventually stopped asking.

This continued through middle school and most of high school. I grew to despise mathematics and hated going to math class every day. It took until 12th grade for a teacher actually to *listen* to me. My Advanced Placement (AP) Calculus teacher realized how much I was struggling. When I informed her that I would be dropping out of the class, she rerouted the conversation and expressed her desire to help me. In fact, she asserted that she would not allow me to drop her course. I told her about my years of struggle and assured her that I did not belong in AP Calculus. "Again, I will not sign your drop slip," she said. "I will help you, but you need to meet me halfway." *Why does she think she can help me?* I thought. Something about her persistence piqued my interest.

She tutored me every day, both before and after school as well as during her lunch period. She was patient, yet firm and clear in her expectations of me. She saw how real my anxiety was. At first, I would have tantrums and resist her help to divert her attention from the task at hand. Patiently, her face never wavering from its determined look, she would wait for me to finish complaining, and then say, "Try the next one."

Slowly but surely, I improved. What happened was truly amazing—I learned the mathematics I never grasped in grades K–11 while simultaneously learning calculus. For the first time in my life, I entertained the idea of taking a mathematics course in college. I was overwhelmed by the power of the transformative relationship that can be formed between a teacher and a student. I thought that I would be cheating myself if I didn't take Calculus II in college, just to see whether

I could make it through a mathematics course without my former AP Calculus teacher by my side.

My freshman year of college was tumultuous; I could not overcome the feelings of homesickness that I experienced. I did take and pass Calculus II, but I was so emotionally distracted that I did not get much out of it and my performance was lackluster. I decided to transfer to Queens College, much closer to home. Taking a tremendous leap of faith—certainly not believing I had acquired enough skill to ever become a mathematics teacher—I inquired about the special mathematics teacher program at the college.

I received an email reply from the director of the program the next morning. "Thank you for contacting me with your goal of becoming a math teacher," she wrote, "… a wonderful goal indeed." She requested to speak with me further. After listening to my story, she said, "Those who have struggled make some of the best teachers," and reassured me that I seemed like a perfect fit for TIME 2000, an undergraduate program for prospective secondary mathematics teachers.

The program was *hard*. I still lacked confidence in my mathematical abilities, but pushed through one course at a time. As the semesters progressed, I improved tremendously. The program faculty was there to support me every step of the way. I started to engage deeply in the program community and took on leadership roles like editing the newsletter and tutoring my classmates in writing. My classmates enjoyed studying mathematics with me because I was so focused on understanding the *why* and the *how* behind the concepts we were learning. Before long, I was tutoring some of my classmates in our mathematics courses as well. My strengths in writing helped me to craft clear and succinct lesson plans which was the key to success in my education courses. I discovered my knack for creativity and sought ways to design fun and interactive mathematics lessons. I grew to treasure every minute of college and was extremely grateful for where my path had led me.

During the first semester of my senior year, I took a course about teaching mathematics using technology. For an assignment, I researched ways to integrate the iPad in mathematics classrooms. The iPad was new technology, and tremendously engaging for secondary students.

My classmates were excited and enthusiastic about trying out the ideas I shared and my professor was impressed with my presentation. She asked me to present my ideas to my peers and many alumni at an upcoming annual program reunion dinner. This meant presenting in front of about 150 people. I had never before addressed a group so large. Although I was nervous, I was comforted by the fact that my peers and professors would be present.

My presentation was successful and once again captivated the audience; everyone was excited about using the newest technology to transform otherwise dull mathematics lessons. I was elated! My professor soon after invited me to present with her at LIMAÇON, a professional mathematics conference on Long Island. She mentored me through the entire experience: We prepared together, practiced the presentation, and refined my talking points. When the day of the conference came, my excitement was accompanied by nonstop nerves. I was presenting in front of veteran teachers and professors of mathematics education. Some of the attendees had been teaching since before I was born. My professor reassured me that I had something new to offer everyone and helped me to stay calm. She opened the presentation by explaining what we would do in the hour-long workshop, and I took it from there. The energy in the room was fantastic; every seat was filled and everyone was excited by the ideas I shared.

At the conclusion of the presentation, we were showered with positive feedback. I felt like a superstar and was so proud of myself and how far I had come. I knew for sure that I was hooked on making workshop presentations, and was eager to do it again. My professor suggested that I "take the presentation on the road" and submit a proposal for the AMTNYS conference. Although I feared taking the leap to make the presentation on my own, I knew at this point that I had grown enough personally and professionally to take on this new challenge. To my delight, this proposal was accepted and the presentation was a success, as described above.

Presently, I have completed six years of teaching secondary mathematics, and three of those years as a mathematics department chairperson. I have given many more conference presentations and workshops. I am currently a mathematics instructional coach, a college-level mathematics instructor, and I am completing a doctoral degree in mathematics education. My journey to the present included many trials and tribulations, as well as triumphs and defining moments. The mentorship I received on my own path to leadership in my field wholly inspired me to empower my own students to grow into leaders. One student's story in particular stands out above all.

During my first year of teaching, I was assigned to teach Algebra I to ninth graders. I worked in a school serving students in grades 6–12, which offered Algebra I to eighth-grade students. This meant that any ninth graders taking Algebra I had failed the course or the Regents exam in middle school, and were taking it again.

On the first day of school in September of 2012, a freshman named Nicole loudly entered my 4th period algebra class. She had neither paper nor pencil, let alone her summer homework assignment, and chatted and laughed with her peers while I tried to get the class under control. She visibly had no desire to be in math class, as evidenced by her behavior and attitude.

The first few weeks dragged by. Each day, Nicole persisted in behaving in ways to disrupt her own learning. She did not wear her school uniform, and in fact, didn't even carry a backpack. Although some of her classmates displayed similar attitudes, she stood out because she seemed to be the most well-liked student in the class.

Moving her seat several times did not stop her from distracting my other students from the lesson, so I decided to call home. I explained to Nicole's mom that she had not completed any homework or classwork and that she was distracting others. Her mom was surprised to hear what I was saying. "She does usually hate math, but she really likes you. She talks about you every day after school—only says positive things." *What? She likes me? She's ruining my class!* I asked her mom for some background information about Nicole's mathematics history and learned that she struggled immensely with confidence in math class. I learned that she had failed the course the year before and had earned a 51% on the Regents exam. After these unfavorable experiences in middle school, Nicole felt like she could not master high school-level mathematics. Nicole was unsure of herself and found it easier to do less than her best rather than put in her all and possibly still face disappointment. Like many teenagers that age, she grappled with a lack of self-confidence and impulsiveness.

Little did I know at the time, that phone call would be the beginning of a transformative experience for Nicole. I stayed awake late that night thinking of myself as a math-anxious ninth grader. I had needed someone to pay me a little extra attention and make me realize that I could succeed, just like Nicole did. I made some edits to my materials for my class the next day, making sure to scaffold some questions with Nicole's skill set in mind.

Nicole entered class the following day and expressed her annoyance with me for calling her mother. I immediately diverted her attention to the day's worksheet and pointed to the problems I had designed with her in mind. "I saw you solve problems like this yesterday. Do you think you could do this one and put it up on the board? Some of your classmates were struggling with this yesterday." Looking at me bewildered, she asked, "Are you crazy?" I persisted, and said that she could bring a buddy with her but that I really needed her to do this. Nicole was hesitant, but seeing that I was not giving up, she reluctantly looked at the problem I had asked her to solve. "OH! I know how to do this," she said, and quickly pulled her friend to the board with her. Afterwards, I asked Nicole to explain her work to the class, and she did so. I couldn't help but notice that Nicole was more attentive in class for the remainder of the period. It was music to my ears when she asked a classmate to borrow a pencil so that she could work on her worksheet.

The next day was the first time that she arrived in class promptly and went to her seat without being asked to sit down. I was encouraged by this step in the right direction and decided to do everything in my power to make Nicole see that she was capable of anything she set her mind to. This included after school tutoring, exam revision assignments, maintaining close contact with her mother, and many encouraging pep talks.

As Nicole started to trust me, she took the first steps towards trusting herself. Changes in her overall behaviors took a long time, and her progress was not linear. Nicole did not consistently complete her homework until the end of the second marking period; however, she was starting to believe in herself and together, we watched her grow one day at a time. She started to raise her hand in class to ask questions and provide answers.

By January, Nicole was walking into math class smiling and with her home-work prepared on a daily basis. Her classmates began requesting to sit at her table so that they could ask her for help. At the end of the year, Nicole scored nearly thirty points higher on her Regents exam in June than her score the year prior. Through patience and perseverance, she was able to see the results of her hard work. Her mother consistently thanked me for helping Nicole recognize her potential, and even wrote a letter to the district superintendent about the impact I had on her daughter. I was excited to see Nicole continue to grow as high school went on.

Nicole spent tenth grade settling into her new self and adjusting to the demands of difficult coursework paired with extracurricular activities and home responsi-bilities. She visited my classroom frequently to conference about her progress. She stayed afloat and did not let frustration get the best of her, but was still not earning the grades that she had come to desire.

Eleventh grade was when Nicole really blossomed. She was ready to shine, and became her own advocate by seeking opportunities for leadership and self-improvement. In the beginning of that year, Nicole stopped by and shared with me that she was thinking of applying for a spot on our school's spring break study abroad program in China. Reflecting on her growth to this point, I was ecstatic that Nicole was considering taking such a large step outside of her comfort zone, where she would be representing our school as a student ambassador, acting as the leader she was meant to be. Nicole applied for a spot on the trip, and to our delight, was accepted.

Upon returning to the United States, Nicole made a presentation about her experiences in front of our entire student body of nearly 700 students. She spoke with confidence and poise. As I witnessed that presentation, I realized that Ni-cole's self-doubt was long gone, and she had grown into the best version of her-self. As Nicole's former persona faded into the background, an eager, self-assured young woman emerged who was ready to make the best of her remaining time in high school.

I was ecstatic to find out that Nicole would be my student again in twelfth grade in a mathematics elective class aimed at preparing students for university-level mathematics. The Nicole who entered my classroom each day as a senior was motivated and dedicated. She walked with her head held high, had clear goals for the future and was unafraid of taking positive risks to advance her own learn-ing. She soared to the top of my class and consistently voiced that math was her favorite subject. As her courses became more challenging, Nicole became more eager to tackle new challenges. Taking her toughest course load yet, including Advanced Placement Psychology, Nicole achieved Honor Roll as a senior and earned an impressive overall average of 89%. The twelfth-grade Nicole was a completely transformed student and individual as compared to the ninth grader I met three years prior. As I eagerly wrote her a letter of recommendation for col-lege the following fall, I was inspired by the capacity for growth and leadership she had demonstrated, and knew she was ready for her next chapter as an undergraduate student.

I keep in touch with Nicole and her mother to this day. Nicole is a college senior, studying psychology with a concentration in child studies. She has become a model student, and has grown into the leader she was meant to be. I am confident that in her future work with children, Nicole will inspire others to be the best they can, by drawing on her own challenges and experiences.

Students like Nicole remind me of why I chose teaching as a profession. She certainly made a lasting impression on me, and helped me realize how possible it was to replicate the mentorship provided to me by my former AP Calculus teacher and my Queens College professor. Teaching mathematics is about so much more than formulas and problem solving. To me, mathematics teaching is about confidence, leadership, and opportunity. Paying forward the mentorship that I received as a student to my students has been the focal point of my career, and has helped me and my students to become leaders, thereby creating the ripple effect that is coveted by educators in every discipline.

CHAPTER 6

LA HISTORIA

Shaded by Violence

Julio Penagos

MY YEARS BEFORE COLLEGE

The first time I witnessed a murder I was 10 years old. I was playing soccer on the street with my friends and we heard multiple gunshots coming from an alley. We quickly ran behind a wall to hide and peek. We saw a man come out of the alley with a gun in his hand, smoke still coming out of the barrel. We then saw how he met another man, and they exchanged money and the gun. It was the early 1990s, in the city of Medellín, Colombia, categorized as the most dangerous city in the world, with around 6,000 homicides per year. This story repeated itself day in and day out for years. It was the norm. So much so, that every night when my

The Inspirational Untold Stories of Secondary Mathematics Teachers, pages 45–52.
Copyright © 2020 by Information Age Publishing

brother or sister went out to hang out with their friends, I would sit down to pray that they would come home alive. I had already lost two uncles and a cousin to this unceasing violence; I knew it could happen to anyone, even to me.

I have had two recurring nightmares. The first nightmare always happened on restless nights with very high fevers. I remember seeing two abstract versions of myself. The first version of myself was a flourishing man, prosperous, organized, and immaculate. The second version of myself was always the image of a muddled individual, dirty, dark, somber, careless. The visions of those two individuals in the nightmare always provoked a lot of sadness and anxiety; I knew those two characters were a version of me. The idea that I was going to become one of those men always troubled me. I am still trying to give meaning to that nightmare. In the second nightmare, a black car parks in front of my house at night and I open the door. As soon as I open it, the car drops a dead body in front of me leaving me paralyzed, and I wake up in horror. Urban violence marked my childhood in a way that I still fail to understand; I suspect it is the reason for that second nightmare.

I grew up in an underprivileged neighborhood in the outskirts of the city. I never felt poor because my parents always made sure I had food, clothes, a roof over my head, and a safe home environment. However, at a very early age, I started to understand the horrors of poverty and economic disparity. Early one morning, a group of hungry-looking children with no shoes or shirts knocked on the door of my house to ask for some bread and sugar water. My mother opened the door and told the kids to wait. She called me and asked me to help her bring some food to them. I was shocked when I saw the children; some of them were my friends. Kids whom I played with during the day had to wake up early in the morning to go knock on doors of houses to ask for some food so that they could survive the rest of the day. That day, I felt privileged always to have food on my table and perturbed by understanding how close I have always been to poverty and inequality.

I lived all my childhood surrounded by poverty and violence, but I still feel I had the best childhood. Most of what could be horrible memories are overshadowed by memories of playing with friends, flying kites, chasing flying bugs and crickets, climbing trees, and spending time with my family.

The first fifteen years of my life passed in front of my eyes full of experiences and contrasts: urban violence and family harmony, social inequities and strong friendships, stories of lack of opportunities and the success story of my father (who rose himself from a poor peasant boy to give his family a middle-class life). Life just passed in front of me, I never felt the need to reflect on my life or experiences; I just lived.

At the age of fifteen I read in a book a quotation from Søren Kierkegaard that stated that "...the thing is to find a truth which is truth *for me*, to find *the idea for which I am willing to live and die*" (1835/2015, p. 19; original in italics). That year my family migrated to the United States. The violence, lack of opportunities, the prospect of my parents' divorce, and social instability motivated my parents to leave our home in Colombia and search for new opportunities in another country, a new beginning.

That year I first met solitude and sadness: I would cry every day for months, being in a place that was not my home, a country with a different flag, different values, a different language, with no friends, and a culture that I could not understand. But I found something in that solitude. I found a place in the sorrow I felt, a place for myself. For the first time, I started looking inside of myself and evaluating my life. The afternoon when I read the quotation by Kierkegaard, it felt like it was written just for me. I started living, searching for that truth that would give meaning to my life.

When we moved to the United States, we traveled with a tourist visa. We stayed, our visas expired, and we spent the next seven years as undocumented immigrants. When you are an undocumented immigrant there are some common feelings that burden you. Sometimes you feel like a criminal with a constant feeling of guilt. Your mind is never at ease, usually feeling inferior to others, many times humiliated, with a constant desire for anonymity. I had all those feelings. However, I also felt life was full of opportunities. I felt optimistic, how could I not? I was living in the capital of the world. I was an immigrant, but I saw, with my own eyes: the dream of a place where any citizen of the world can prosper. That dream was the essence of New York City.

During my first months in America, I became a frequent visitor to a place that changed my life, the library. At fifteen I became a voracious reader. I became obsessed with reading Latin American literature. Gabriel Garcia Marquez, Jorge Luis Borges, and Pablo Neruda became my idols. I decided I wanted to be a writer. The beauty of their literary styles plus falling in love for the first time equated to a passion for writing. One day a teacher in high school asked us a question about what we wanted to be. I said proudly that I wanted to be a writer and change the world with my words. The classmate next to me replied that he wanted to be a teacher. At that time, I remember thinking to myself how I would never in my life be a teacher. It seemed like the most boring, least respected profession of all.

I knew my dreams of being a writer were just dreams. As time went by in high school, I saw it as impossible because I could not write well in English. It was very challenging, still is. Every short story or poem I wrote was in Spanish. I slowly decided that I could not be a writer in a country in which I could not write the spoken language. I also realized that being a writer was not going to pay bills in the future that easily. I had heard of "probability" in mathematics class, and decided that my probability of being prosperous would increase if instead of being a writer, I became an engineer. That became my plan. After all, I felt I was good in my mathematics classes, and I was told engineers were always good at doing mathematics. During my senior year in high school, I applied to local colleges that offered engineering programs. I got accepted to all of them. It was exciting to think I was so close and yet so far from the opportunity of going to college. I was accepted to all those colleges, but I knew I had no money to go to school, and as an undocumented immigrant I had no access to financial aid. I knew I had to finish high school, get a job, probably low paying, save as much as I could, and then go to school, maybe for a semester, and then repeat the process every year until I graduated. That was my real plan. Going straight to school to be an engineer was more of my dream plan, an impossible one.

My math teacher during my senior year in high school was a light in my life. I always thought it was amazing that he was a mathematics teacher and his last name was the same as the number five in Spanish. I thought life made him to be a math teacher. He probably did not have to look too hard to find his truth, it was always written in his name. One morning, I went to his office to ask him for advice about college. I will always wonder what force led me there, nobody had sent me there, and I was not the kind of student who would ask for advice from teachers. I sat in his office and asked him about the schools to which I had been accepted. I inquired which one was the best for engineering. He gave me his choice. Then I asked him whether it was too expensive. He quoted some average amount. Then, even though I never wanted anyone to know I was an undocumented immigrant, I told him of my situation. I explained to him how I had no money and no documents to go to school, and how I was not going to be able to go because of my financial and legal situation. He smiled at me and asked: "Do you want to be a teacher?" "No," I said. "I never really thought about it," I added. He then said, "If you want to be a math teacher, there is a scholarship to which you can apply." I repeated to him that I was undocumented and that was out of my reach. He told me I should still think about it, and if I decided I wanted to teach math, to apply.

That day, after school, I took a long walk, thought about my life, reflected about all the things that I had experienced before getting to that moment. I thought it was too good to be true, to be able to go to school for free while I lacked legal status in the country. It seemed impossible. But what other choices did I have? This was my only opportunity. I had to try something. I decided to apply. If by some miracle, I got accepted, I would complete a math teacher degree, get a job as a teacher, and then pay for more schooling to be an engineer and become one. It seemed like a sound plan.

The next day I arrived at my teacher's office and told him I wanted to do it. I applied to the TIME 2000 program, a four-year college program that paid full tuition for a degree in secondary mathematics education. He guided me through the process and I sent the application. I was anxious for many days, thinking it was too good to be true, thinking it was not going to happen. I thought to myself, with resignation, that it was going to be much harder for me to get a professional degree. I would probably start working and end up like most people I knew that were in my situation, working for the rest of their lives, with a low paying job, because becoming a professional seemed too far out of their reach.

The day I received the letter of acceptance to the scholarship program, was one of the happiest days of my life. With tears in my eyes, I was able to tell my family that I was going to be able to go to college on a scholarship. I was able to relieve my parents of the obligation to make ends meet to help me pay for school. It was a happy day.

FORGING A PROFESSION

After my first day of college I knew I had landed into something special. I had heard horror stories of large lecture halls, with hundreds of students, with mean professors who did not care about their students. My first class was the opposite: a small room on the bottom floor of a strange maze-like building. That first class had about twenty students, and a professor who would become one of my most important mentors, models, and heroes.

During my years in college I discovered that the dark individual from my first nightmare was always present, it was part of my personality: chaotic, careless, and almost rebellious. I had a feeling that this alter ego was going to prevent me from graduating. In the midst of that chaos I met very dear and valuable friends. In my TIME 2000 classes, seminars, and events, I met people from almost every continent. We had the world in a small classroom in Queens. I have the privilege of saying I made friends from the United States, Romania, Guinea, China, India, Ecuador, Puerto Rico, Dominican Republic, Japan, and many other places. These friends deserve more than a paragraph in this story. They were there to remind me of a very dear lesson I had learned from my mother: always to surround myself with friends and to believe in friendship. My friends were a support during those college years. Because of the structure of the scholarship program, we moved as a group, took the same classes together, the same study hours, and the same school events; we moved as one. Some of them were good with deadlines, some of them were very organized, others took the best notes (they were essential at the time of studying for mathematics exams), and others were insightful, charismatic, and humorous. My flaws were minimized because we worked as a team, we complemented each other. I am forever thankful to my college friends. Because of them I have a career.

As my years in college progressed, something peculiar started happening to me. I started falling in love with the subject, with mathematics. I suddenly no longer wanted to be an engineer. Mathematical ideas started captivating my mind and soul. Because of the quality of some of my professors, I learned to see beauty in mathematics. After becoming obsessed with mathematics, I realized I was not like most people. I had a favorite number, like most people, but I also had a favorite equation, a favorite theorem, a favorite mathematician, and a favorite geometric shape. This love for mathematics was in a sense, a symbol for my identity. Mathematics was like me, like my recurring nightmare, the one about the two versions of me. There were two sides of mathematics, the practical one that involved laws and theorems, difficult to understand, full of order and rules; it involved lots of discipline and mastering it was almost painful. The other part of mathematics was all about beautiful patterns and relationships, equations that defined the movement of the planets, or the attraction between subatomic particles.

This part of mathematics was artistic. Einstein said it better: "Pure mathematics is, in its way, the poetry of logical ideas" (1935, p. 12). I learned to see both sides of mathematics and learned that they complemented each other; they were two sides of the same coin. Maybe mathematics was just a metaphor for who I was. The two individuals in my nightmare could have just been me all along. Somehow, I had to embrace it. My years in the program instilled in me the belief that mathematics is in itself a form of art, a language, and a science. And that it should be taught with the strictest rigor, with clamorous joy, and with perpetual passion.

At the beginning of my senior year of college I was still undocumented. I often asked myself about the purpose of going to school and getting a degree since without legal status I was not going to be able to work. All that hard work was not going to give me the reward I wanted at that time: to teach in New York City public schools. I just wanted to teach mathematics and show the world how amazing it is. Moreover, having been exposed to all the poverty and injustice I saw growing up, I wanted to make a difference in this world, to make it better for everyone.

One day, around the middle of that school year, my mom asked me to accompany her to a meeting she had with a man. She told me that this man was my father's boss, and he had the power to change our lives forever. My parents had been trying, for many years to do everything they could to change our legal status in the United States. After lots and lots of paperwork, countless hours spent with immigration lawyers, and days of tears and desperation, it all came down to this man's signature. He had been helping us throughout the whole process, but because of misunderstandings with my father, he decided that he was not going to go along with it. In fact, he decided he wasn't going to sign the last paper. That afternoon, my mother and I entered the man's house. I remember the house seemed small, muted, and with very little sunlight. When the man came out to the living room, my mother started pleading with him to sign the paper. The man insisted that he had made up his mind, that he was sorry, that he was not going to sign the paper, and that she was wasting her time. Then I saw my mother do something I will never forget. She got down on her knees with tears in her eyes, begging the man to sign the paper, not to do it for her or my father, but to do it for her children. The man signed the paper. A few months later, before graduating from college, I received my green card, a piece of paper that gave me the ability to teach in New York City. It will take many lifetimes for me to pay my mother back for all of the sacrifices she made for my family and for me.

PURSUING A DREAM

"Why are you here?" asked the student.

"I am here to help you as much as I can and teach you math," I replied.

"You are not going to help us," she replied. "Nobody cares about us. All of the teachers in this school leave. You don't care about us. We are not listening to you."

That was my first day of teaching, right after I introduced myself to the students. The student was probably right. I did not know how to help her or her classmates. How was I going to help students whose mothers were sick, or without a job, or who lived in a shelter? How could I help students whose parents were alcoholics, drug addicts, or were in jail? Some of them did not even have parents. I had no idea. That first year of teaching, I did not know what I was doing. I was so well prepared to teach math after going through my undergraduate program. However, this was a different ball game. Not even the best mathematics education program can prepare you to teach in one of the most dangerous and underprivileged neighborhoods in New York City. I tried different things every day to teach my students, to get them to listen to me, or attempt to do any work. I often failed. I was trying to help my students with different lessons and strategies every day, with little to no success. I came into teaching thinking I was going to change the world in one day. After a few days I realized how wrong I was. I realized after a few months that it takes a long time and great sacrifice to create change. I realized I needed to do something good for these students, and I found the perfect tool to do it: teaching mathematics. At the end of that year, the student who challenged me on the first day had become the best student in the class; she told me that my class was her favorite.

The school in which I worked my first two years of teaching was located between two of the most dangerous housing projects in Brooklyn. It is difficult to feel safe in a place like that since students often came to school with knives and daggers. There were frequent student fights, sometimes shootings outside the building, and teachers' cars being stolen. One day, a colleague of mine told me about a shooting right outside the building, and bullets had hit another teacher's car. The teacher was not in the car, but the car ended up with a lot of bullet holes. I looked at him and dismissed it as a normal occurrence. That night I remember thinking about the whole situation and reading about it in the news. Seeing the event in the news made me realize this was not a normal occurrence, at least not for most people in the United States, but it was for me. Growing up amidst so much violence numbed me to these types of events. The horrible experiences of violence in my childhood had given me the strength I needed to teach inner city adolescents. I was able to understand their narratives, their actions, and their lives better.

Children in deep urban areas of New York City grow up seeing crime and violence as everyday events; they have to grow a thick skin because safety is not guaranteed. They go through life being discriminated against because of the color of their skin, their ethnicities, and their social status. I came to realize that these children whom I was teaching were just like me at that age, with similar fears, similar struggles, and the same sense of uncertainty about their futures. I ended up sharing my stories with them, showing them that besides being their math teacher, I could also be a mentor, because I learned to feel their struggle. I realized I could teach them better when I listened to them, understanding why they could not come to school because they had to care for their siblings when their parents were working, or when their parents were too sick to care for the little ones, or when their parents were just not there. Listening to them helped me understand why they would come to school late, high, or afraid. Unexpectedly, teaching and doing mathematics with my students became easier when I understood them and their struggles. Misbehavior was always there, but I understood it, so it stopped creating issues in my class. I became a better teacher because I learned to be vulnerable with my students, to listen to them, and more importantly, when I saw myself reflected in their stories.

The first two years of teaching were transforming for me. The first day I entered the room with a degree in mathematics education. Two years later, with all the work I did, and all that I had experienced teaching in that school, I felt like I had achieved a degree in how to learn about and meet the diverse needs of the complex human components of adolescents.

After a few years, I accepted a teaching position closer to my home. My new school had a different vibe. Students came from all over the world, with different cultural backgrounds and languages. I continued to search for a truth to guide my life and my actions, as Kierkegaard suggested.

At the end of my twelfth year of teaching, I went through a period of deep reflection. I reflected on my career and all the lives I was able to touch. I reflected on my passions, my visions, and my dreams. I reflected on years I spent studying mathematics education. I realized my truth started to come to light during those years; I developed a deep passion for mathematics thanks to my experiences with my professors, my mentors, and my classmates. The years of preparation in TIME 2000 planted the seeds and some very strong beliefs: the belief that students learn best when they learn together; the belief that students learn best when teachers enjoy the beauty of the subject they are teaching; the belief that high quality teaching is the best predictor of student success; and the belief that teaching is an inspirational and praiseworthy profession. My truth is my destiny: mathematics teaching and mathematics learning.

CHAPTER 7

GOOD-BYE SHYNESS,
HELLO TEACHING

Michael London

It is Friday afternoon after a long week of teaching and I am excited for a social gathering with friends at a local restaurant, eating my favorite comfort foods such as chicken wings, French fries, and pizza. As we all gather, we exchange pleasantries as we have known each other for many years and just as we are about to take our first bite of our delicious food we begin to chat about our greatest passion in life, mathematics teaching. As the conversation starts I secretly ponder how I got to a point where I would be so excited on a Friday afternoon to gather with a group of math teachers and combine fried food with fractions.

Growing up, one would say I was a shy and introverted person who would often limit communication to a few close friends and family members. When I would think about what I would like to do in the future I thought of such careers

The Inspirational Untold Stories of Secondary Mathematics Teachers, pages 53–57.
Copyright © 2020 by Information Age Publishing

as being a mail carrier or a train operator. I could have never envisioned myself being responsible for the mathematics education of over one hundred students every day. As I went through my childhood I developed a passion for numbers and statistics. My father and I would often play "math" games involving making fractions to represent the situations happening in our lives. For example, I remember us sitting around the table having pizza and after having our first slices my dad would ask how much remains. Furthermore, growing up I had a great passion for sports and would always be checking the baseball standings and statistics to see how many games the New York Mets were out of first place. Unfortunately, after a brief statistical analysis I would often find out that the probability of them making up this deficit was very small.

Based on these mathematical experiences, as I entered high school I had a passion for numbers and statistics. However, teaching was far from my mind in that I still lacked the personality for it. Furthermore, believe it or not, in my first two years of high school, mathematics was not my best subject. The turning point of my mathematical career as a student came during the first month of my junior year in high school. I was enrolled in an Algebra 2 honors class. One day I saw my previous year's math teacher speaking to my current math teacher. I was called outside of the class for a private conversation in which I was told I had not met the requirements to be in Algebra 2 honors and that I would be removed from this class. This was the low point of my time as a student of mathematics. I felt like a failure in that I wasn't able to be good in a subject that I was so passionate about. From that day on I decided that mathematics would no longer be a weakness and that I was going to become a great mathematics student. After being transferred to my new class I began working very hard, completing extra problems, going for tutoring and I began to see that mathematics was not that difficult. I realized that if you understood what you were doing then you could go from being a weak student to a strong student in a short period of time.

In my rise from the ashes as a mathematics student, I was very fortunate to have some great teachers. It was at this point I began to see how important having a good mathematics teacher is. Towards the end of my junior year I got to experience a slight taste of what it was like to be a mathematics teacher in the form of peer tutoring. Being a member of the National Honor Society, I was assigned to tutor a freshman in algebra. Having never done this before I was a bit nervous. However, as time went by and I saw my student was improving, I felt a great sense of satisfaction, a feeling that I wanted to be replicated in the future. This tutoring experience also made me realize that learning the universal language of mathematics could allow me to communicate with anyone and thus help me overcome my introversion.

Entering my senior year of high school, the ingredients to become a mathematics teacher were beginning to come together. My passion for mathematics was combined with academic success and I had begun to help others and communicate mathematically. It was towards the end of senior year that everything came together. I had been accepted to Queens College and was then given the pamphlet that would change my life from that day forward. The pamphlet was an advertisement for the TIME 2000 program at Queens College. The program was a scholarship program, which prepared members to become secondary mathematics teachers. In reading this, I felt like this could be the career that would combine my passion for mathematics with the satisfaction of helping others. When I was accepted into this program, I felt so excited to have the opportunity to one day be able to help others overcome their difficulties in mathematics just as my high school teachers had done for me.

During my first few years in the program I began to see a whole new side of mathematics teaching. I had always thought of a math teacher as someone who just stood in the front of the room and lectured to a silent class. An experience that changed how I viewed a mathematics teacher occurred when the TIME 2000 program provided the opportunity to observe a very special math teacher during my first semester. The teacher had the students talking in groups and she was always smiling. She spoke to them with respect as if they were her colleagues. As I sat in the back of this classroom I thought to myself: This is the type of teacher that I would like to become.

The more I went through the program the more excited I was about becoming a mathematics teacher. We were learning so many strategies to keep students engaged and motivated. Moreover, we were given many valuable experiences outside of the classroom. For example, we had opportunities to meet and hear guest speakers every month and were given the chance to attend conferences that were designed for mathematics teachers. Because of these experiences, I began to realize that I was part of a community that would last for a very long time. In fact, each year current and former mathematics teachers from the entire mathematics education program gather together for a reunion which is a night where during dinner individuals with experience varying from no teaching experience to over 30 years of service come together to socialize and tell their stories about teaching. This is more than just a reunion of math teachers, it is a family excited to see one another. This was an invigorating experience, which provided a strong motivation to complete the program to become a mathematics teacher.

At this point, I had the passion for teaching, the motivation to become part of a community, and the will to help people. However, I still had to face my fear of being in front of a group. In my final years of the program we were given the opportunity to teach multiple lessons in front of a class of our peers. Since we were teaching students we had known for years it did not seem as intimidating. I felt like I belonged in front of a classroom and helping others. In completing the program, I had a sense of pride and confidence that through the language of mathematics I could communicate to everyone.

Entering the field as a mathematics teacher can be very challenging and intimidating. However, as a result of my college experiences I did not feel alone. I had become close with many members of the mathematics teaching community who were just a text away at any time. This amazing sense of community has remained throughout my eight years of teaching.

I was offered my first teaching job at a school where all of the mathematics teachers as well as the assistant principal had been members of the TIME 2000 program. It was amazing that members of the program whom I had gotten to know for the past four years would now be my colleagues. The community in which I had spent four years during my undergraduate studies would now continue into my teaching career. I couldn't be working with a better group of individuals. I felt comfortable. I felt at home.

During my first year of teaching I faced the challenge of how to keep a class engaged and focused throughout the entire lesson. It was often the case that while I was teaching, the students would be talking. In order to overcome this challenge, I turned to one of my colleagues whom I had met through the undergraduate program and therefore felt very comfortable asking him for help. I told him I was having trouble keeping my geometry students engaged while I was teaching proofs. It was at this point that he suggested that I observe his geometry class. As I sat in the back of his classroom, I noticed that as he was teaching, the students were very focused, taking notes, and were on task.

After observing his class, I realized one of the issues I was having was that I was spending time going over problems the students had already been able to solve on their own. The main difference between my colleague's lesson and mine was that he was very strategic in how he spent his time in front of the class. He would only go over the parts of the proof with which he saw the students were having trouble. Therefore, there was a lot more time in the lesson for the students to be working on problems and less time with him in front of the room. This experience was very valuable in that I learned the importance of observing student progress before continuing with a task and deciding what to review as a group. Having the ability freely to observe someone I had known for a long time helped me learn a valuable lesson that I constantly use in my teaching.

The mathematics teaching community I was in when I first entered the TIME 2000 program now extends beyond my own school. Ironically, since my professors recognized my ability to speak well in front of large groups of people, I have had the opportunity to present at various teaching conferences over the years. At these events it gives me such pleasure to see other former members of the program whom I have known for almost a decade. It is truly like running into members of your family. We are there to greet each other with smiles and hugs and to wish each other good luck in our presentations. I am now a member of the Math for America master teaching program, which provides wonderful professional development. It is very often that I will see other teachers I met through the TIME 2000 community. It has been wonderful having the opportunity to know these people from the time they were teenagers to now when they are seasoned teachers. We have all come so far together and we will all be there to support one another for the remainder of our careers.

As I have described, in my experience in order to learn mathematics it is very important to feel like you are part of a community. Therefore, in my classroom I create a community in which the students feel comfortable asking each other questions and explaining mathematical concepts to one another. I do this by creating a system in which every student's response is valued. The system I use to create this atmosphere is that the students are often given a problem to work on individually for a few minutes and then are asked to share their responses in groups. In order to create equity within the groups, I make sure that everyone has a chance to speak before any communication takes place between members. This is done by assigning each member of the group a specific number from one to four so that students can be randomly selected to speak first. As a result of implementing these norms the students feel very comfortable working with one another and they feel as though their responses are valued. After implementing this structure for a few weeks, the students effectively communicate mathematically and value each other's ideas. This community setting can be very helpful to students, who like myself, may have excellent ideas but have trouble expressing those ideas to their classmates. For example, I had a student, Veda, with special needs who would always write excellent explanations and score very well on assessments. However, she would keep all of her ideas to herself and remain quiet throughout the class. Veda had a disability that made it difficult for her to pick up on social cues as to when to begin speaking and how to obtain the attention of her classmates. This community structure greatly benefited Veda in that she had time to formulate her thoughts during individual work so that she knew exactly what to say when it was her turn to speak. It was wonderful to see her begin to communicate with her group members and as a result the entire group's performance on assessments improved. Once the other members of the group noticed how helpful Veda could be, they would often ask her to explain concepts to them. Consequentially, this community helped Veda go from feeling left out to being an extremely helpful resource to her classmates.

As I take my first bite of the chicken wing about to share a meal with members of my mathematical family, I raise a glass to all we have accomplished and all that we will be able to celebrate together in the future. In my mind I think to myself: I have evolved from a shy and timid student to a veteran math teacher surrounded by wonderful colleagues. The value of a community of learners is just as important for students as it is for teachers. I have found a family that pervades my career and contributes to my ability to excel and contribute to the profession. I try to create an atmosphere in my class so that my students can find a family in their fellow classmates and excel in mathematics.

CHAPTER 8

TUTORING TO TEACHING
AND BACK AGAIN

Nerline Payen

GRADUATION DAY, JUNE 1998

Graduating from kindergarten was a big accomplishment for me. I was proud of myself for learning how to read, write certain words, and for knowing (or at least thinking I knew) what I wanted to do for a living. My kindergarten teachers told us that when our name is recited at graduation to take a few moments to share with the audience what we wanted to be when we grow up. My classmates stated the common ideal professions of a 5-year-old: basketball player, movie star, actress, musician, and so on. When my name was called, I marched up proudly to the microphone and stated, "When I grow up, I want to be a teacher." Where did that idea come from? Who knows?

The Inspirational Untold Stories of Secondary Mathematics Teachers, pages 59–65.
Copyright © 2020 by Information Age Publishing

Maybe it was because I had such great kindergarten teachers? Maybe it stemmed from both my parents being in the education profession? My mom shows me this tape recording of myself every year and I always think to myself, it sure is funny how life comes full circle sometimes. One may think that I always had my mind set on being a teacher, but there were definitely some bumps along the road that led me to where I am today.

FAST FORWARD: 2018

It was a Sunday night and the doorbell rang. It was a friend of my husband who came over to spend some time with him, and I was up in my bedroom getting things ready for work the next day. I was upstairs, but I could still hear them talking between themselves downstairs. My husband's friend was telling my husband about some online trivia game that he found out about that allowed players to win money if they answered all the questions correctly. The game was going to be live in a few minutes and he wanted my husband to witness how the game was played. "Hurry up and go get Nerline to play," my husband's friend requested. "Why?" my husband asked. "She's upstairs getting ready for work." "Come on we need her to win" the friend insisted. "What makes you say that?" My husband questioned. "She's a math teacher man, she's smart." This is the stereotype I've grown up with the past few years. My friends, colleagues, classmates, and family members would always come up to me and ask me for help with all types of problems. And I'm not just talking math problems...I mean ALL problems. They assume that because I enjoy mathematics it means that I am good at everything. They would come to me with challenging math problems and brain teasers to assist them. Sometimes, I am able to answer them on the spot, other times I would say I would think about it get back to them. When I give the latter response, I always get puzzled looks. They would wonder why I couldn't figure out problems on the spot and would ask, "Aren't you gifted? Doesn't it come naturally to you? Isn't that why you became a math teacher?" They assumed that one had to be gifted in order to be good at math and that if you didn't have that gift or if you weren't born with that "privilege" then it was impossible to enjoy mathematics.

I grew up in a culture where the "ideal" professions are doctors, lawyers, and engineers. My older brother has a severe case of autism and was not able to function normally, so my parents put a lot of pressure on me to be the "star" child of the family. Being the star child of the family meant getting good grades, getting a scholarship to college, and taking on one of the "ideal" professions. I wanted to please my parents, but I also wanted to make sure I was happy, so it was my goal to find a good balance between the two.

Once I entered middle school and was given the opportunity to take more science classes, my dad started pushing the medical field into my mind. "Nerline, I think you would make a great doctor! You are very smart, and I think you would enjoy it and they make a lot of money. Imagine, you would be the first doctor in the family!" Hmmm...this idea sounded convincing. Making a ton of money and making my parents proud? In grade 8, I was offered the opportunity to take a pre-nursing course. In this class we learned about the human body, different health concerns, medications, and how to respond in certain life-threatening situations. We dissected animals, took CPR training courses, and visited hospitals in the area. It didn't take long for me to realize that the idea of blood freaked me out as I could barely dissect a frog. I tried to force myself to enjoy the class but in reality, I knew that this wasn't something I could see myself doing. I loved the idea of helping those in need, but the medical field just wasn't my calling.

I gave up on the dream (or should I say my dad's dream) of becoming a doctor and decided it just wasn't for me. One day I was having a conversation with my little brother which turned into a discussion, which eventually became a heated debate. This happened often with us and I always like to proclaim myself the winner of the debate, at least that is what seemed most logical since my brother would always give up. My mom would say to me, "Oh boy, the way you love to have discussions, you should be a lawyer. You are great at having debates and you could be the first lawyer in the family!" Again, this idea sounded convincing. Maybe I could make a great lawyer? I tried out for the mock trial team during my first year of high school and though it was interesting, I couldn't find a passion for it. I liked helping my debate team members come up with viable debate arguments, but I didn't see law school as something I wanted to pursue. I soon said goodbye to mom's dream as well.

Throughout high school, I took various mathematics classes and had the privilege of having great math teachers. I really enjoyed solving problems and getting to a solution. It didn't take long before my friends classified me as the weirdo for grasping the concepts so quickly and enjoying it. "You really like math?" my friends would ask. "You must be one of those gifted kids or something." That is when the idea of being "gifted" started to come into my head. I didn't feel gifted, I just liked math. I liked numbers. I liked brain teasers. I was developing a passion for mathematics. I loved helping my friends prepare for a test and they would always tell me that I should become a teacher. I would always smile at the idea of that but always kept that thought in the back of my mind. As the years went on and I took more mathematics courses, I developed more of a love for mathematics and wondered whether there was a profession that could allow me to explore this for a living. The thought of being a teacher that was stuck in the back of my mind started making its way to the front. I approached my parents for suggestions and let them know that I loved math and wanted to do something with it for a living. I had never seen my parents smile so hard. Their faces both lit up and my face lit up with them. Were they also going to suggest that I go into teaching? I felt that if they said this, it would give me the ultimate validation to move forward with it. My parents opened their mouths to speak and I was almost certain of what they were going to say. "You should be an engineer!" they both blurted out. "Imagine you would be the first engineer in the family!" Engineer? That hadn't crossed my mind at all, but I guess it was worth looking into. I didn't mention the idea of teaching at all because I didn't want to disappoint them. I decided just to move forward and see how the next few months would go.

In junior year I started looking into colleges to apply to. I looked at schools with engineering programs, but also looked to make sure those schools had teaching programs as well. I came across a flyer for a program called TIME 2000 at Queens College, a program dedicated to preparing secondary mathematics teachers. I had to put the flyer down and pick it up again to make sure I was reading it correctly because it seemed too good to be true. A program that combined both teaching and mathematics? I felt like this was the sign I was looking for. Reading the description of this program made me realize that being a math teacher was my passion because I felt excited. I tried to force myself to be excited and enthusiastic while reading engineering program descriptions, but I just couldn't feel it. Being the cautious person I am, I applied to several schools and programs, but I definitely had a first choice.

In the spring of senior year, I started receiving responses from different schools. "Congratulations! You have been accepted…" they started. My parents were ecstatic and I was happy, too, but I was still waiting to hear from the Queens College program. A few weeks later I received the letter, an acceptance letter from the program and I felt a huge sense of relief. This is my sign! I was over the moon but at the same time worried about what my parents would say. I showed them the letter and my dad asked, "You want to be a math teacher? Why didn't you ever tell us that?" I was so concerned about being the "star" child and being the first in the family to do something that I didn't even consider approaching them with the idea. My dad seemed happy for me, but my mom was quiet the whole time. What was she thinking? Why wasn't she saying anything? She left the room, walked into her bedroom to get something and came back to the living room with a DVD. She popped the DVD into the DVD player and a video began playing. It was my kindergarten graduation video. Why is she playing this? I wondered to myself. She fast forwarded the clip to the part where I went on the stage to recite my name and my desired profession. "Hi, my name is Nerline and when I grow up I want to be a teacher." She laughed out loud and blurted, "Isn't this funny?!" I couldn't help but smile to know that I had the support of my parents to pursue my dream.

I began the program in the fall of 2010, and immediately knew that the program would be demanding and challenging, so I knew that I would need to be devoting enough time to focus on my studies. When I began freshman year, I had an afterschool job that continued from the summer, which didn't pay much per hour, but was a way for me to make some money. It was a job working with young children which was great, but with the long hours and little pay I knew it was going to interfere with my schoolwork. I decided that it would be best to leave that job and just focus on school, but I definitely felt the impact of that on my wallet. It was tough being a college student, yet alone a broke college student.

After leaving my afterschool job in the fall of 2010, I had a little more free time and wanted to get more involved with the TIME 2000 program. Being in a cohort with my classmates was great, but I was seeking a way to interact with more TIME 2000 students and invest more time in the program. One day I met with the program director, and I asked her how she thought I could contribute to and get involved in the program. She recommended becoming a part of the TIME 2000 tutoring club and at that moment that idea felt like the perfect solution. I could make money in less time and gain teaching experience at the same time? Win-win. The program is run by TIME 2000 students who interact with parents, manage tutoring jobs, and assign them to other students in the program. In my sophomore year at Queens College, I became the vice president of the tutoring club and in my junior and part of my senior year, I became the president. I learned so much by running this tutoring club. I was able to get the best of both worlds by being involved in the program and making money through tutoring jobs. Being a tutor gave me even more validation that I wanted to teach.

I will never forget my first tutoring job, a set of third-grade twins. One would think that since they were twins, they would be of similar levels mathematically. Wrong. One twin was extremely advanced and loved mathematics, while the other struggled tremendously with grasping concepts. They were SO different. I was introduced to the idea of differentiation before I even knew what differentiation was because I had to approach working with and reaching these two students in different ways. I knew that the advanced twin enjoyed mathematics, but I was worried about the impact I had on the other. A few days before Christmas, each twin gave me a card. I opened the first card, from the one who loved math. *"Dear Nerline, You are a great tutor. Thank you for challenging me and making me love math even more!"* A huge smile came across my face. I then opened the next card, from the other twin. I must admit, I was a little nervous to open this one. *"Dear Nerline, Thank you for being patient with me when no one else is and for being my tutor. I learned a lot from you and this year I'm good with math. I hope you will always be my tutor."* A small tear rolled down my cheek. I've never felt that I made more of a difference in my life and this gave me even further validation that I wanted to teach. Until this day, I am still tutoring these twins and they are now going to high school. They invited me to their middle school graduation, and I felt like a proud parent. I felt like a difference maker.

My name is Nerline Payen and I currently teach at a middle school in Queens, New York. The school is one of the largest middle schools in the district, housing approximately 1,500 students. It is a home to a diverse student population serving students from several different backgrounds and cultures. This is my fourth year teaching at this school and I am very happy to be a part of this community. The school's mission is in line with my teaching philosophy of focusing on a student-centered environment, preparing all learners with differentiated curriculum that is relevant, meaningful, and engaging to students. I've had great moments in my teaching career which have reminded me why I wanted to teach, but I have also had some challenging moments that I've had to overcome.

One challenge occurred in my third year of teaching when I taught one young girl of a set of twins. I have had a lot of experience teaching sets of twins, but this year I had a particularly hard time differentiating between the two. They were identical and I mean IDENTICAL and they could easily fool their teachers and friends. Their personalities seemed very similar as well, down to their sense of humor, interests and overall demeanors. I had one of the twins in my general education sixth-grade class and the other twin was in an honors class. At the first parent teacher conference meeting in November, I met with the parent of the twin I had in my class who was also struggling tremendously in mathematics. Her sister was also at the conference but sat at a different table, so I was able to differentiate between them. Whether she had sat at a different table or not, I would have still been able to differentiate between them at that moment. It was the first time I noticed a distinct difference between the two girls. One seemed confident and one seemed defeated. We sat down at the conference and her mom immediately asked her, "Why do you feel like you don't do well in math?" "I don't have the math gene mom, you know that; I'm just not gifted like that," she responded. There goes the idea of being gifted again. I interjected for a moment, "What is the math gene?" "Oh, it's when you're born with the ability to do well in math. My sister has it, but I don't," she responded. "Is there an ELA gene or a Social Studies gene?" I asked. The twins looked at each other and laughed. I noticed the mom laughing, too. Why was this such a crazy question to ask?

Each year I've encountered several students who feel defeated. There are those who feel like math is a subject they cannot and will never be able to grasp, and there are others who feel that they just don't have the "gene." In my first year of teaching I went into the profession and said to myself, "I'm going to make sure every single student I teach feels like they can succeed in mathematics." Reaching all of my students has probably been my biggest challenge. There have been times that I have prepared a lesson that I was sure was going to reach every single student and by the end I would notice puzzled looks on a few faces. This really bothered me because I didn't want anyone to fall behind and there were some students I felt like I just wasn't connecting to, especially those who required individual attention. I would hold extra help math sessions during lunch and after school, but there was still only one of me and many of them. I thought about how much more my tutees benefited from having one-on-one attention and an idea popped into my head. One day, I asked my honors students if any of them would like to come to join me for lunch or afterschool extra help sessions and tutor some students from my other classes who were struggling. Before asking, I debated what I could use as rewards for this. Maybe I could give students some extra class participation points? Maybe I could count this towards community service hours? Maybe just some lollipops and stickers would do it? Once I asked the class I was shocked to see how many hands went up. I was expecting two, maybe three students, but they were all willing to help without even hearing about any possible rewards. I then thought to myself, why limit this to my honors class? I gave the option to my other classes as well and many students were willing to help. I organized a schedule during lunch periods and afterschool sessions and the next thing I knew, all of my students were getting small group and individual help. This helped not only build friendships and relationships among my students, but also helped to build confidence within my students. Many of my helpers would report back to me that they felt like they understood the material even better now that they were able to teach it to someone else. They felt like leaders. The students who were receiving the help felt grateful to have peers be patient with them and work through concepts with them. I would just oversee the process and answer questions as needed, but for the most part, my students ran the show. I even took tips from my students on how to teach certain concepts. The only thing I could ask myself is why hadn't I thought of this before?

I must admit that reaching the goal of meeting the needs of all students has been harder than I thought. There are still times that I am so frustrated with myself when I have students who are just not improving. But then there are the times where I do see improvement, passion, and a love for mathematics develop within my students. At times like this, I remember why I wanted to teach.

Looking back, people always tell me that I always knew that I wanted to be a math teacher but in reality, I didn't. Teaching had to find me. How did I know it was meant for me? I just knew.

CHAPTER 9

FACULTY SUPPORT GOES A LONG WAY

Young Mee Kim

Did your teacher ever ask you whether she could adopt you as her daughter? Well, that was what my professor asked me. It was my senior year of college when many of my peers were busy interviewing and receiving job offers as mathematics teachers in the secondary schools. My professor connected me with opportunities for interviews as well, but my situation did not permit me to accept the job offers that came after successful interviews; my immigration status hindered me from working. My professor did everything that she could to find a resolution for my situation, even seeking out legal advice; however, she realized an unbreakable wall was set before me.

The Inspirational Untold Stories of Secondary Mathematics Teachers, pages 67–70.
Copyright © 2020 by Information Age Publishing
All rights of reproduction in any form reserved.

My name is Young Mee Kim and I hold a full-time assistant professor position at an urban college in Flushing, New York. How could this have happened since that memorable day in my professor's office?

When I was in elementary school, I wanted to be a teacher, but negative opinions from others around me persuaded me to turn my eyes toward professions with greater monetary returns that many high-achieving students consider. However, my immigration status limited my choices for college and my path was deterred to the TIME 2000 program, a teacher preparation program offered by a local city college. The program rekindled my passion for mathematics and reminded me of my calling to be a teacher. Upon graduation, I could not wait to share all of the amazing things that I learned in the program. I was fired up and ready to go, but I was held back by my legal status that did not permit me to work. I was over the age of twenty-one when my parents gained their legal alien status, so I had to reapply as a child of permanent residents over the age of twenty-one. Due to my age, I was not able to gain legal status and all I could do was to wait until my application was processed. I was deadlocked.

It was around this time that I had the earlier conversation with my professor. I had these frequent meetings with the professors in the program. We were asked to write journals at least once a month, and we even held small group conferences with the professors to share our experiences in the program and any difficulties that we encountered. The professors knew about what was happening to us and discussed among themselves areas of concern. It was difficult for an introvert like me to share my deepest agonies with anyone, but interestingly, I found myself talking and sharing my problems and concerns with my professors. The frequent meetings and reflections facilitated the sharing process.

It seemed as though the only open door was the path to post-graduate education until the situation was resolved, so I continued to obtain a master's degree. I enjoyed and excelled in the graduate-level courses and in educational research, and it led me to wonder about further educational opportunities. Again, I consulted my professors to decide my next step. They directed me to a graduate school with doctoral studies in mathematics education. I did not know or dream of attaining a doctoral degree. Coming from a family where I was the first person to complete a bachelor's degree, a doctoral degree was beyond any of my expectations. Nevertheless, I continued to study and put my utmost efforts in my work, instead of focusing on the circumstances that I could not fix.

Near the end of completing my doctorate, I stumbled on another bump in the road. The qualitative research that I was conducting required eight participants fitting specific criteria; an ambitious condition to begin with. After all, who would want to spare their precious time with barely any payment in return to participate in a research study conducted by a graduate student? They would be providing service to the educational community out of the generosity of their hearts. No wonder I was at a stalemate again. I ran to my professors for help. At this point, I was no longer their student in the program, and they had no obligation to help me, let alone spare time to meet with me. Well, yes, as you have guessed, they found participants for my study and I was able to resume my research.

My legal status was adjusted in the last year before the completion of my doctorate, and finally I was able to share my passion in teaching mathematics. After weeks of sharing the resolution of my situation, I received multiple e-mails from my professor wanting me to interview for a part-time teaching position at an urban college. I went to the interview and was offered a part-time teaching position. I still remember the first class that I taught. I was thrilled and exhilarated that I was able to put all my knowledge and preparation into practice. I was thankful for each class, no matter how unmotivated or uninterested in mathematics the students might have been. I was still able to share the joy and beauty of mathematics. It was my time to help my students appreciate mathematics. When the full-time position opened, I was instantly offered the job.

In serving my students, I attempted to get to know them, as my professors had gotten to know me. Learning and understanding students' backgrounds, needs, and interests became my mission. I created and administered a questionnaire to determine students' interests, hobbies, career paths, attitudes toward mathematics, and any concerns they had. This information was considered when creating cooperative learning groups as well as in developing lesson plans related to students' areas of interests. Some of the freshmen students were grouped according to their majors to facilitate the process of adjusting to the new college environment. Applications of the mathematical topics from students' majors and interests were discussed in class and students were asked to work on mathematical projects related to their discipline of study.

Further attempts in learning about the students were made through monitoring student progress as well as by having conversations with them. Students and group collaboration were monitored, and students were asked to write brief notes about their classroom experiences and their attitudes toward the group work. Non-functional groups were identified through this effort. Once I was puzzled by students who requested to be in the same group not working well with each other. I found out from the brief notes that the students submitted that the group members had arguments outside of the class which hindered their group collaboration. The group was altered to encourage a positive learning environment.

I tried to meet with the students and talk to them before and after class and sometimes in the hallways to find out about their learning experiences and areas of concern. If students were struggling, I reached out to them. I strive to create an engaging, supportive, and positive learning environment. Furthermore, students in my class were asked to complete an anonymous written evaluation of classroom practices at the end of the semester inquiring them about the positive aspects and areas for improvement related to their classroom experience. The student written evaluation along with the student evaluation surveys (i.e., Student Instructional Report II (SIR II)) conducted by Educational Testing Service, an independent third party, were considered when making future plans to improve instructional strategies.

Once I noticed a hard-working student beginning to show tardiness and even missing classes. I found out from my discussion with the student that he recently began to hold night shifts at an additional job along with the two other positions that he held. These employments were necessary for him to pay for school and to provide for his living expenses. I spoke to the student's advisor but there was not much that I could do to help him. I felt helpless because it seemed as though the words of encouragement and the listening ears were all that I had to offer. I told the student not to give up and that there surely was a light at the end of the tunnel, reflecting on my situation. Then I turned to what I could do which was to make my instruction meaningful and exciting. The student had a big smile and sparkling eyes in class. Ultimately, the student was successful in the course and continued toward the completion of his degree. Whenever I meet the student in the college hallways, he still greets me with a big smile.

At another time there was a student in my elementary-level mathematics class who came back to pursue a degree in engineering after years of not studying mathematics. The student seemed to be afraid and puzzled at first with the new endeavor that he took on, but slowly, he was guided back to mathematics. The co-operative learning groups formed in class permitted him to make new friends and often he was found studying with his group members. The student worked hard. I made opportunities to listen to students' concerns in and out of class, where at times the student shared difficulties he was facing at home and hindrances he encountered in improving his academic achievement. He was encouraged to persist. Slowly, the student got back on track. I noticed a positive change in his attitude towards mathematics. This student, who was much older than I, often greeted me with a big hug and addressed me as his "math mother."

In one of the interviews that I had to get a teaching position, I was asked about what I considered to be most important in teaching. I was taken aback by a question that I had not prepared for, but after a minute or so, words from my heart said, "Preparation and compassion." I pondered about that response. Preparation, I understand, but why did I say compassion? Mulling over this response led me back to the compassion exemplified by my professors. The word compassion was deeply rooted in me due to the genuine and heartfelt concern and support that I received and now I am paying it forward to my students.

CHAPTER 10

CHALLENGES, SURPRISES, AND SUCCESSES

Kendal Askins

It was my second of year teaching and we were having parent-teacher conferences. As I welcomed the parents and showed them where to sit, I noticed that the focus was not on the student we were there to discuss. The parents were looking at me, questioning my understanding of the material that I was teaching their child. The mom kept referring to her husband as "the smart one" and stating how much he understands mathematics and she doesn't. As I tried to redirect the conversation, the husband stopped me and challenged me with a riddle that he put on my desk. "Here figure this out!" Annoyed, I looked down; I had seen it before. I had a flashback and could see myself sitting in my undergraduate program's monthly seminar. I could hear the presenter's voice, how could 2=1, it can't! I knew the answer and I answered the question with no hesitation. The proof in-

The Inspirational Untold Stories of Secondary Mathematics Teachers, pages 71–75.

volved the illegal operation of dividing both sides of the equation by zero. I could see eyes light up when I knew the answer; there was a new level of respect. "Is she right? Is she right?" asked the wife as the husband sat there amazed, replying, "How did you get the answer so fast?" I could now continue with my parent-teacher conference with a big smile across my face. I was a teacher who really understood mathematics in this parent's eyes. I understood the material and was now a part of a "secret club." This parent was now willing to listen to me and not only respected my opinion but me as a professional.

My name is Kendal Askins. After being in the profession for 12 years, I have had my share of challenges, surprises, and successes throughout my career. The first 6 years I taught in a junior high school in New York City. After moving to Pennsylvania (PA), I was a gifted-support teacher and I helped provide services for gifted students in elementary through high school. Currently, I am the assistant principal of an intermediate school in PA.

When I accepted the position as a public school gifted-support teacher, I had a picture of what my students would be like. I felt that I would be working with students who loved school—Wrong! Always did what they were told—Wrong! Never had discipline issues—Wrong! Could handle high-level material despite the subject—Wrong! Taking this position helped me to see that I really knew nothing about gifted students, and that is where my learning began. I had been in the classroom for six years at this point and had felt very comfortable with average-ability students. My experiences have allowed me to deal with special education students, English language learners, honor students, and average-ability students. I thought that a gifted student was an honors student and I began this journey with that in mind.

Being a gifted-support teacher challenged me to be better and to re-evaluate my thoughts and what I had become comfortable with. I can recall one situation with a student who would never do his homework. Homework was about twenty percent of his grade and he just would not consistently complete it. What was surprising to me about this student is he would get 100 or high 90s on all of his exams. The exams that were developed tested high-level concepts, not just recall information. This student had a B+ average in the class solely because he would not complete all of his homework assignments. I decided to have a conversation with him. At the time it was my belief that it was important for every student to complete the assigned homework. When speaking to the student, I learned that he felt the homework was a waste of his time. Before reacting, I continued to listen. He explained that he understood the concept and could demonstrate his understanding, so why should he have to sit and do different problems that repeat the same concept? He said that he liked the word problems and he would review all of the questions every night and choose which ones he felt he needed to do. It was at this time that I really understood the type of students I was dealing with. I was not working with students who were not aware of what they could handle academically. I was dealing with students who did not want to waste their time but rather consume their time with new ideas and challenges. I responded to this student by stating that he had given a fair assessment of his behaviors and together we worked on a plan that would benefit him.

After this experience, I realized that there were other teachers who worked with these students who had similar biases that I initially had. I began to take opportunities to work with other staff members to help them see what I was seeing and to help them gain a new understanding of their gifted students. There were many myths and misunderstandings among the teachers about gifted students.

Using these new insights, I had to think quickly of meaningful ways to motivate my gifted students so that they could meet their high potentials. I was challenged to think of activities for them that were appealing to me as a student in my undergraduate teacher preparation program: portfolios, real-life applications, seminars, and trips! I set them all in action!

PORTFOLIOS

I was proud of the portfolios that I created in college and while I used them in my mathematics classroom, I wanted to figure out how to incorporate them in a meaningful way in my gifted classroom. I wanted my students to understand themselves: their abilities, interests, learning styles. Creating their own portfolios would be a means to accomplish this.

Every month the students were responsible for collecting at least three samples of their work for their portfolios. These samples had to represent either an interest, ability, or something about their learning style. Furthermore, they had to write a rationale for each piece describing why they chose to include it in their portfolio. The students had total creative control over how they organized their portfolios. One year a student created a chess game. He made the pieces and a huge wooden chess board by hand. Each piece represented a decision he had made that school year. One side represented the good choices and another side represented the bad choices. He even planned out all of the moves, and used rationales to explain the moves the pieces were making. During his presentation he "played the game" reading his rationales as he moved his pieces.

I also had a student who loved to read and draw. She decided to create a collection of books as her portfolio. There were three books during her first year of creating her portfolio. One book represented her interest, the other her learning style, and the last, her ability. She made the books hollow and included her artifacts inside. Her rationale was in the form of a story that continued throughout each book. During this student's second year she chose to add books to her collection, expanding her story and using that to describe her growth.

There were no portfolios that were alike. Some students used technology and created movies, slide shows, websites, and so on. The common theme was that they were exploring who they were, how they were growing, and what they were good at. The students were learning so much and enjoyed creating a product that represented themselves and their development as learners of mathematics.

SEMINARS

I knew that I needed time with my students outside of their academic classes. I created seminars that were divided by grade level. I met with the different grade levels to discuss not only the developments of their portfolios, but other higher level, real-life application topics that they would not have had the opportunity to be exposed to in their on-level academic classes. Every seminar would begin with a problem-solving activity. I would put the problem on the board and set the timer. The students enjoyed working on these problems because they were challenging and different from any of the problems they had seen before. At these seminars, I also had the students reflecting on their classroom experiences and communicating what their needs were. From these conversations it became clear that students wanted an understanding of how the material they were learning could and would be used in real-life. From this need another program developed.

REAL-LIFE APPLICATIONS

The students craved more time with real-life applications. They were learning these concepts in school, but it wasn't school that they enjoyed. It was the learning and the challenge of understanding how to use the information that they learned that they enjoyed. The program that the teachers of the gifted students and I developed gave them this opportunity. We called this program Intermediate IF[1] Institute. This program took place monthly and focused on high-level concepts that would challenge and relate to what they were learning. There were speakers, projects, and opportunities to help expose them to concepts that they may have known existed but never truly understood.

One lesson that I designed and taught was about fractals. The students had never heard of fractals before. We spoke about the mathematics behind fractals and where fractals are found in nature, technology, and mathematics. The students also had an opportunity to build both a two-dimensional and a three-dimensional replica of Sierpinski's Triangle. This was engaging for the students because they learned about the mathematics behind Sierpinski's Triangle and they were using that knowledge to construct the replicas.

Another lesson that was conducted was about "simple machines." Students not only learned what simple machines are, but they were able to explore multiple cases by working at different stations throughout the classroom. One station required them to create their own "Rube Goldberg Machine" (i.e., a complicated machine that performs a simple task), using a variety of simple machines. Another station required them to create their own machine that would draw a star on a piece of paper. The students enjoyed these stations because they had creative control over the machines that they built.

---

[1] IF refers to "What If?"

TRIPS

As I continued to get to know the students, I realized that many of their experiences were limited to those that they received in school. I also realized that many of the students only had the limited experiences that the teacher had. Every year they went on the same local trips. Coming from New York City, I knew everything that New York had to offer, and I wanted my students to have exposure to that as well. While I understood that it was important for the students to know what their community had to offer, it was also important that they had experiences that went beyond their local community.

Their favorite trip that we went on was to the MoMath Museum in New York City. The students had never been to a museum like this and many of the concepts that they explored at the museum were concepts that they had been introduced to through our monthly time together during Intermediate IF Institute. It was thrilling to see their excitement during the museum visit. They called me over to show me how the different exhibits worked and how they experimented with them. They also made connections with some of the concepts that we had discussed together in class. Many of the students who did not enjoy the idea of school were excited to participate in Intermediate IF Institute because it fulfilled their craving for learning and applying knowledge. After this trip, the students had an opportunity to write a reflection about their experiences at the MoMath Museum. They shared how grateful they were for the experience and how much they learned. They were so appreciative and were looking forward to visiting the museum again.

Being a gifted-support teacher challenged me as an educator. It also allowed me to take the mathematics that I love and bring it to life for my students. Through the portfolios, seminars, and trips my students had an opportunity to develop as people and experience real-life applications of the mathematics they were learning in their classrooms. Through their reflections they have expressed the positive impact that these experiences have had on them. Even though I have moved on to administration, the seminars and the Intermediate IF Institute still continue.

Being an educator is not just a job for me, it is who I am. I have wanted to teach since I could speak and my love for teaching continues to develop every day. I believe that one person can make a difference and I strive to be that person.

CHAPTER 11

I COULD ONLY IMAGINE

Sabrina Joseph

I was going away for college, far away. I was going to dorm and love it. I was all set, with my college t-shirt and keychains for my parents, because I WAS GOING AWAY FOR SCHOOL. At least that was what I thought. Making plans for where you will go to college and what you can actually afford are two *very* different things. I quickly came to realize this when there wasn't any money for me actually to go to school out of town.

Once it became clear that I was going to be commuting to the college fifteen minutes away from the house I had grown up in, I decided to make a long shot phone call and see whether the honors program for aspiring mathematics teachers, TIME 2000, had any space left. After all, it was June. The application process was over, and the orientation for incoming freshman had already taken place; but I thought, I might as well give it a try. When I called, a kind lady with a sweet voice (i.e., the secretary) answered the phone, and her response was

The Inspirational Untold Stories of Secondary Mathematics Teachers, pages 77–80.
Copyright © 2020 by Information Age Publishing

"we will always make room for a good student." She had no idea how those words would change my life—and I could only imagine. When I hung up the phone, I quickly gathered the documents for my application packet and hand delivered them to the office so that it wouldn't waste any time. And then, I waited.

Several weeks later, on a hot summer day while walking with some friends, I missed a call from an unknown number. The caller left a voicemail message. I remember saying "Oh my gosh! Oh my gosh!" while listening and my friends looking at me slightly concerned asked, "What happened?!" What happened was that I was in! I was offered a seat in the program.

Grades came in for the first semester of my first year in the program. Everything looked FANTASTIC, except for that C- in calculus. Not only was a C- terrible, but it also meant that I was on probation in the program. It meant that I could lose my seat and the scholarship that went with it. My first thought was to run. You are NOT cut out for this. The self-doubt crept in quickly. I asked myself, "Did you *really* think you could be a math major?" If you can't handle the first semester, how will you handle the next three and a half years? I was sure I should quit, but my mentor, a senior in the program, was sure that I should persist. He encouraged me to keep going. I guess program administrators knew that while going through the coursework you might need someone who has been through it to guide you. Dee was that person for me. Anytime I felt that I wasn't good enough, or smart enough, or just that I wasn't enough, he assured me that I was *more* than enough. He became my rock during college and has been ever since. Twelve years after first meeting, we have been married for 10 years and have two sweet children.

Before I knew it, I was graduating and looking for my first teaching job even though there was a hiring freeze in New York City. It was a difficult time since permanent positions weren't available to new teachers. I was committed to getting my foot in the door somewhere, so I interviewed for a substitute position for a teacher who would be out on maternity leave for six weeks. I went in feeling confident, and by the time I got in the car after the interview, I shut the door and sobbed. I was POSITIVE that I had completely blown the interview. I didn't even bother turning my phone on. Hours later—after a long nap—I turned my phone on and BAM, there was a voicemail from the assistant principal with whom I had interviewed, stating that he would like to offer me the job, and that he didn't know why he didn't just offer it to me on the spot. WHAT?! I was shocked and thrilled. My first job! Those six weeks went quickly, and I continued my search for a permanent position.

Finally, in February 2011, I was hired in the middle of the school year for a permanent position at a school in Manhattan teaching eighth-grade mathematics. The principal at the time had to jump through a lot of hoops to hire me since the freeze on new hires had not yet been lifted. I was thankful, and I am still thankful that he took a chance on me, a beginning teacher. Early on, a more senior teacher and I were talking, and she said with a smile and seemingly confident in my abilities, "Well you know all this stuff, this isn't your first-year teaching," to which

I replied shyly, "It is." She was so surprised to hear that, because when she observed my lesson and my interaction with students she was certain that I had more experience than I actually had. I attribute this to the preparation that I received in my undergraduate program. I was able to observe in classrooms starting in the very first year of college and learn from other professionals at math teacher conferences that were just for teachers—even though I was not yet a teacher. These opportunities that were afforded to me in my undergraduate program helped to mold me into the teacher that I would become.

My philosophy of education can be summed up in my six-word mantra: "Teach like they are your own." My students—whether they are the ones in my class or just the ones I know from seeing them in the halls during passing—all deserve to have someone in our school building in their corner. With close to 4000 inner-city students and around 250 faculty and staff in our building in Queens—all on over lapping schedules, I believe there should be a way to get this done. For many students, this means more than we will ever know—and sometimes as teachers, we are blessed to see a piece of how much the relationships we build with our students matter.

After the high school graduation, at the end of my fifth year of teaching even though most of the faculty had left, I stayed behind to look for my students—one in particular, Sonny. I was so proud of him. Getting to graduation was not an easy road for him as he had to overcome many obstacles to make it to that point. As the area cleared, I finally saw him standing with a friend looking around. I went over to congratulate him. Sonny met me with a huge grin, a big hug, and told me he was still there because he was looking for me. As he put it, I was his "second mother" who pushed him to cross the finish line. He said he couldn't have done it without me. These words (in a letter he had given me) brought tears to my eyes. As an educator, it is my firm and unwavering belief that these kinds of supportive relationships are essential for all students.

Now, having been a teacher for nine years, with the last seven in the same school, I have had multiple professional and leadership opportunities. For example, I have had many student observers and student teachers. I have given presentations at professional conferences for high school students interested in becoming teachers. Most recently I have built a partnership with my alma mater hosting TIME 2000 sophomores and their professor in my class. The professional network created by the program has given me a platform to communicate with and contribute to the next generation of teachers about the importance of building supportive relationships with students. In my mind, such relationships coupled with strong content and effective teaching methods, have the potential to change a student's life. If the job of an educator is to prepare our children for the future, then how can anything be more important than teaching them, treating them and loving them like a family member?

About a year ago, as I reflected on why I became a teacher and why I love to teach, it became clear to me that teaching the 150 students I have each year isn't enough. There are so many more students who need to be reached and staying in my classroom isn't going to do it. That was when I decided that I would take the step into school leadership—something I swore I would never do. When I started teaching, it was with the vision that so many new teachers have: I want to make a difference in the lives of my students. Now I want to make a difference in the lives of as many students as I can. I intend to do that with the teachers I will lead. It is a great challenge that I embrace wholeheartedly.

As an 18-year-old with her dreams of going to college out of town being crushed, I would have never thought that this would be where I am 13 years later. Sometimes doors get shut, or slammed, in your face and you feel like you are failing, but as we have all heard before, setbacks are just set-ups for epic comebacks. I believe that it was always all part of God's plan for my life—to shape me and to help me always to be grateful. I look back now and am so thankful that I didn't go away to college and more thankful for the kind lady with the sweet voice who answered the phone and told me that there would always be room for a good student.

CHAPTER 12

THE POWER OF CARING

Alexandria Capozzoli

STORY #1: MY STARFISH

Where to begin? I'm not sure if every teacher has "this one student." I've never thought to ask my many teacher friends. Regardless, I do. I have "this one student" that when I think about *why* I do what I do...when I think about *why* I spend hours lesson planning and grading, *why* I drive 2 hours each way commuting, *why* I report to work the day after a student tells me he "hates math" or *why* I update my online grade book from midnight to 2 am when I have to be back up at 5 AM, for some reason *this one student* always casts a positive light over my thoughts. *He* is the reason. Whomever *he* may be. I just want to make a difference.

I'll never forget my first year of teaching. I don't think any teacher ever will. The first year is the hardest. Things only get easier from there. For me, it was September

The Inspirational Untold Stories of Secondary Mathematics Teachers, pages 81–88.

of 2011. I taught seventh-grade mathematics in a middle school in Long Island City, New York. The demographic was… how do I say, "tough"? Many of our students came from the local projects. Almost all of our students were under-performing. They were told year after year that "You are a 2" or "You are a 1." What did my co-workers mean when they said things like "Well, don't bother with him. He is a 1"? Doesn't everyone *want* to be Number 1? Isn't "Number 1" a good thing? Well not when my co-workers are referring to the students' test scores on the state exam the previous year. I had 3 sections of seventh grade that year. I had my honors class. It was the class that was on track to take the algebra Regents exam in eighth grade. "It is your high class. The students are pretty much all 3s and 4s," I was told. OK… I wish they were described differently, but that's how it went here, I guess. Then I had my second class. It was a general education class. The students were set up to take the algebra Regents exam at the end of ninth grade. Instead of describing this class as "on-grade-level," I was told, "you'll have some tough kids behaviorally in there. But don't worry, they're mostly 2s and 3s. Most of them will do the work you tell them to do, at least *some* of the time." Wait a minute… *most of them* will do their work? Why wouldn't *all of them* do their work? They'll do their work *some of the time*? Why not *all the time*? OK let's not think about that because they're already explaining my third class to me. I hear "… So basically they're all 1s and 2s. You won't get any teaching done. Just try your best to keep them settled down behaviorally and minimize write-ups to the dean. Since it's an integrated co-teaching (ICT) class, they're split into 2 sections. You'll have half the students in your classroom and the other teacher will have the other half of the students in her room. You'll each teach half of the students with individualized education programs (IEPs). Luckily, they won't all be in the same room. Any questions?"

My brain is screaming:

WAIT, WAIT, WAIT. I have 592,635,926 questions!

First, why are we referring to the students as 1s, 2s, and 3s?

Second, what do you mean they don't do their work?

Third, I won't get any teaching done? What will I do for 43 minutes and *not* get any teaching done?

Now the questions are flowing through my head faster than I can filter them.

"No, no questions," I said with a smile. "Thanks for the information! I can't wait to meet them tomorrow." I wasn't lying. I couldn't wait to meet them. But the only thing on my mind was how we were told in my undergraduate program four short months ago, that other teachers would describe students and put negative ideas into our heads. Never listen! Allow your students to prove themselves in your classroom. Well, thank goodness I was prepared for this. If not, I probably would've been dreading the first day.

So why did I give you this background? It's important to set the stage so you understand about "this one student." I will tell you that he was "a 2." He was a "low 2," meaning that if I didn't pay close enough attention, he could score lower this year and get a "1." In the beginning of the year, we had to identify students who we wanted to be our "focus students" in each class. We'd spend extra time helping these students and ensuring their success. We were told the best way to choose our "focus students" was to "choose pushables or slipables." They told us this, but quickly realized they'd have to explain what that meant considering neither of these words could be found in the dictionary. "Pushables" were students who had a score of 1.9 or higher, which would make it easy to "push" them to a 2, a score of 2.9 or higher which would make it easy to "push" them to a 3, or a score of 3.9 or higher which would make it easy to "push" them to a 4. "Slipables" were students who had scores barely over 2, 3 or 4, which made it easy for them to "slip" to the next lower score. The reason we wanted to push students up to the next score, or prevent them from dropping to the lower score was to meet Adequate Yearly Progress (AYP). Meeting AYP looks good for your school, so it was something the administration focused on.

I did as I was told and chose my "pushables" and "slipables" to form my focus students, but I just wasn't sure why we weren't just aiming to *push ALL* our students to the next level. Anyway, upon looking at the data, before knowing my students too well, I chose Leo because he obtained a score of 2.26 on his state test last year after getting a score of 2.91 the previous year. He seemed to be a nice boy with a positive attitude. I figured he would be a good candidate to work with because I didn't want him to continue to decline. Although, I'm not going to lie; I wasn't planning on trying to keep him at a level 2. A score of 2 is not proficient. My goal was for him to get a score of 3; or better yet, a 4. So, Leo was one of my 3 focus students in my lowest level class. The school year began, and it was my mission to get to know my focus students the most: learn what makes them tick—both academically and otherwise. Leo was a real kind and genuine kid. He was so likeable. He was friendly with *all* his classmates. This should be easy.

In order to practice the math skill, the students learned each day, we have the students work in groups to complete their "work period." Back then, their work period would consist of them applying their skill to make a foldable (i.e., a three-dimensional, student-made, graphic organizer) or other creative application. I noticed a trend in a lot of Leo's work. He often drew or referenced crowns. I never thought anything of it. One time, I put his work up on the bulletin board outside my classroom because I was so proud of his hard work. One of the other students pulled me aside and told me he was surprised I'd hang it. I asked him why. He explained that the symbol he drew on the coordinate plane was not just *any* crown. It was the symbol that represented the Latin Kings and that Leo was a member. The minute this student was out of sight, of course I ran to my bulletin board and took down his work! I couldn't be displaying this inappropriate gang reference in the school hallways! Some teachers may decide that their time would be better spent on a different student after finding out this student was in a gang. For me, it was *quite* the opposite. Suddenly, thinking of him as a "pushable" was the furthest thing from my mind. Now my reason for choosing him as a "focus student" was to change his *life path*. He was too good for this! Well, all kids are—but *especially* him. Not only will I get him to improve in math, I'll get him to improve his life. I have to. It's my mission. If I accomplish nothing else this year, please let me accomplish this.

Luckily, I managed to support my students and provide an environment where I had virtually all my students complete, or at least attempt their work *every* day. Work period time was anywhere from 10 minutes to 30 minutes depending on whether there was a single period or double period. This was a great time for me to walk around and help students or groups who were struggling. While helping students, this was also a great time to get to know them better and have small talk in between the math. I always found that the more you know about your students and the more they feel you can relate to them, the better they perform, because they feel that you care about them and they want to please you. As often as I could, I would make small talk with the students at Leo's table. I didn't neglect the other students, but I made sure to get to know Leo the best I could. In doing so, I found out that he wanted to be in the service when he graduated. He was in the seventh grade and he was adamant that this was his life goal. When I asked him "why?" his response was "to get out of this kind of life." He explained that things weren't easy in the area where he lived. He told me that he wanted to do better but didn't think he could do it while living there. Of course, in the back of my mind I wondered whether he *knew.* Did he know that Rodrigo told me that he was in a gang? I didn't ask and he didn't give away too much information. I left the conversation at that. Every time that any future discussion lent itself to it, Leo talked about being in the service. He wanted to be a Navy Seal. Often times, he would draw the symbol for Navy Seals or mark up his paper with the word NAVY repeated infinitely. The more I worked with Leo's group, the more I saw all their quiz and unit test scores increasing. This was promising. At the end of the year, I looked back in my grade book and saw that Leo went from getting a 65 average in the first quarter, to earning 70, 75, and then finishing with an 80 average in the final quarter. I was so proud of him. I hoped he was happy! I know he surely seemed so. I was just hoping I had left a lasting impression.

Much to my luck, the following year I found out that I was moving to teach eighth grade. The only downside was that I was changing academies. Each floor in our school was a different "academy." The idea was to break the school into smaller schools so that the students felt that they were in a smaller, more inclusive environment where the teachers and administrators really knew every one of them. I was nervous to be in the "other academy" where I didn't know the other teachers on that floor, *and* I didn't know these students. Surprise! I had another ICT class that would be split with another teacher. When I was given the list of students, so many sounded familiar! Many of my students from Leo's class were transferred downstairs in order to provide them with the ICT class setting that was necessary based on their IEPs. I was happy to read that Leo was among the students who I would have in my half of the class for this school year. I couldn't wait to continue to help these students with whom I had been working so hard.

To make a long story shorter, I worked closely with Leo again this year to make sure he could continue to progress and improve his mathematics scores. His quarter averages in math (and across the board) continued to climb. In the first quarter of eighth grade he started in the low 80s and finished the year in the high 90s! What an improvement! Over the two years I taught him, he went from being on the border of failing, to one of the highest performing students in the class. He was now graduating middle school with amazing grades in all subjects and he was accepted into a wonderful high school! I questioned him as to why he chose *this* school though. I asked him "why would you choose a school that specializes in this and not a school for students who plan to join the service?" "Miss," he responded, "I am not going to enlist anymore. That's what I wanted to do last year when I didn't think I had any other option. Now, that's not good enough for me. I can do more. I'm going to be a successful career-driven adult." WOW! My heart melted and exploded with excitement all at the same time. Whether or not I had any influence on this positive attitude he was showing me, I know that his hard work was going to get him far and even if I'm not solely responsible, I was so happy to be along for the ride.

At the end of the year, I won a Teacher Recognition award from Children First Network (CFN) 410, which was my school's network. At the award presentation, the awardees were asked to read a poem named "The Starfish Story" (see Table 12.1). We were asked to reflect on our teaching experiences and think about whether there was a particular student for whom we made a difference. I'll never know if my impact on Leo was as profound as the impact he had on me, but he gave me a sense of worth. He made me feel that if I really tried, I could help my students. And no, I can't work miracles for every student... but if I could pro-

---

TABLE 12.1. The Starfish Story*

One day a man was walking along the beach when he noticed a boy hurriedly picking up and gently throwing things into the ocean.

Approaching the boy, he asked, "Young man what are you doing?"

The boy replied, "Throwing starfish back into the ocean. The surf is up, and the tide is going out. If I don't throw them back, they'll die."

The man laughed to himself and said, "Don't you realize there are miles and miles of beach and hundreds of starfish? You can't make any difference!"

After listening politely, the boy bent down, picked up another starfish, and threw it into the surf.

Then smiling at the man, he said, "I made a difference to that one."

*Adapted from "The Star Thrower," by Loren Eiseley (1964)

foundly help just one student my whole life, let alone one a year, my life would have purpose. The poem entitled, "The Starfish Story" explains exactly how I feel about Leo, my Starfish (Eiseley, 1964). When you're teaching, if things ever get tough, read this poem and remember not to let anyone tell you that there are too many starfish to help. Make a difference, even if it's only for "that one."

Leo graduated middle school in June 2013. He started high school in September 2013, and was on track to graduate high school in June 2017. I thought of him often, especially when things were tough, and I questioned why I do what I do. I did wonder whether he remembered me and whether he was living up to the high expectations he set for himself. It was unfortunate that students who graduated were not allowed to come back to our school and visit their former teachers. I figured I'd never know what happened to Leo. Luckily for me, Leo proved me wrong. He sent his cousin to see me during the 2015–2016 school year. She attended the middle school where I taught. I'll never forget the first day she knocked on my door during one of my prep periods. "Are you Ms. Kubic?" "Yes, what can I do for you?" I wondered whether she was sent to deliver a message from another teacher. Her faint voice asked, "Do you remember Leo Molina?" "Of course, I do! Do you know him?" "He's my cousin. He sent me to find you. He said to say 'Hi' and tell you that he's doing great!" I was so happy to hear this. I sent her back with a message. The same message I tell all my students when they graduate "Make me proud." I couldn't be prouder to know that in June 2017, Leo walked across the stage to obtain his high school diploma. This is why we do what we do!

STORY #2: A GUT-WRENCHING HALF DAY

It was a Monday in November. Only 12 of 30 students handed in their homework. This was unacceptable, but I needed to know *why* they didn't do their assignment. Was there some issue with how I taught the material and the students weren't able to apply what they learned? Was there a special event this weekend that prevented so many of them from completing their homework? Was it a coincidence that so many students didn't complete it? I needed to know. The best way to find out, was to put a halt on the Do Now that was on the board and ask the students to take out a piece of paper for a short journal entry. I asked them to write about the following: If they did their homework, how or where did they do it and whether or not they received any help with it. If they did not do their homework, explain why not.

Leslie almost always did her homework, so when I got to her journal, I was surprised to read: "I didn't do it because I have a lot going on. My family has problems. I have trouble with integers. At home, I do my homework in my room, on my bed." I read her journal that night. I made a mental note to find her on Tuesday and just make sure everything was OK. Usually, when students write something like that, they want to share more.

The bell rang to end PM homeroom. It was a half day, so it was 11:20 AM, and the students were filing out to go home. It was the Tuesday of Parent-Teacher Conference Night. We would have 2 sessions. We'd see parents from 12:30–3:30

and again from 5:00–8:00. I couldn't wait to see whose parents would show up. I was eager to walk the students out. I always enjoyed getting to talk one-on-one with the parents so I could learn more about my students. Many times, the parents who showed up were those whom I previously had difficulty contacting.

I saw Leslie in the hallway on her way out toward the exit. I told her I read her journal and I wanted to know if she wanted to talk to me about anything. She stayed back as the rest of the class filed out the door. We were standing in the hallway. She started to tell me that there were issues at home. I asked her what was going on and whether she wanted to talk about it.

"My dad hates me," I thought she said.

"What?" I wondered what she meant by that.

"My dad hits me ... he beats me." Now I realized I misheard her last statement. She said, "my dad *hits* me" and that hit me like a ton of bricks! I thought to myself, "OK, Alex *think*. They told you what to do when you were in TIME 2000. Remember, *you are a mandated reporter.*" So now everything she told me, I would have to report to the State. I remember being hyper aware of everything she was saying and paying close attention so that I could recall as much of what she told me as possible.

Leslie continued by elaborating. She explained that her father was extremely abusive. She said that sometimes she wouldn't do anything wrong and her dad would come upstairs from working at the store he owned, conveniently below their apartment, and would beat her. She said sometimes it would be so bad that she couldn't walk back to her room, so she'd crawl to her room. She said that she didn't know why he would beat her because she couldn't think of anything she had done to deserve it. All she knew was that if he came in with an angry face, it would be bad. I wanted to scream "nothing you could have possibly done would have made you deserve that!" but I had to keep my cool.

Leslie continued by telling me that she didn't want him to get in trouble because he speaks English and her mother doesn't. She said that she wouldn't be able to pay the bills and her mom wouldn't be able to because she doesn't speak English. Leslie was worried that if her dad got taken away, that the bills wouldn't get paid. After that, I asked if she had any brothers or sisters. She said she was an only child and that's why she needed her dad—because no one else would be able to help her mom. Then, Leslie told me again that it gets really bad sometimes when her dad beats her. She continued by rolling up her sleeve and showing me her arm. I looked at it—it had scratches all up and down. It looked like scratches, not cuts. She said "this is what I do to myself. I hurt myself to deal with the fact that my dad beats me" and she continued, "I'm hurting."

At the end of our conversation, I told her that although I said I wouldn't tell anyone, this was a serious situation. I reassured her that I would NOT tell her dad, but that I had to tell someone because this wasn't a safe situation and I didn't want her getting hurt. I told her I would tell Ms. Yowozl, her counselor, and she said okay.

I was so calm, cool, and collected on the outside, but I was raging on the inside. I couldn't believe what this poor child was dealing with. I couldn't believe that she was hurting herself to deal with the hurt that her father was inflicting upon her. I was breaking a sweat, trying to remember the protocol for this type of situation. I was so disheveled. I immediately went to her counselor's office. It was everyone's lunch time. I told the counselor it was an emergency and gave her a brief description. She told me she had to eat and would deal with it later. I get it, they see a lot of these issues and maybe it desensitized her a little but that was just *not* acceptable to me. So, I went upstairs to the other school counselor. Maybe she wasn't Leslie's counselor but at least I could get another opinion. I felt terrible that Leslie even walked out of the building and went home to her dad. If there was anything we could do, I wanted it done as soon as possible. Luckily, Ms. Sambol felt as passionate as I did. She called Child Protective Services (CPS) right away. We began the process. It was grueling, but I felt better knowing it was reported. Now it was up to them. At least I did my part.

CLOSING COMMENTS

What do these two stories mean and why are they so important to me? When people ask me what is the *most important* aspect of classroom management—and more importantly, being a teacher, I always say the students need to know you CARE. Caring is not something you can learn, but you can learn *how to show* students that you care. Everything I have done in my teaching career to show my students that I care about them and that they matter has been in ways that I learned in my undergraduate teacher preparation program and by talking to other program graduates and colleagues for their opinions on how to handle situations. Students need to know that you care about their learning, but more importantly about *them.* And the best way for students to know you care about them is by forming relationships with them. Even something as simple as noticing their hair cut or their new shoes or a fancy outfit that they're wearing that day or even just asking them how they're doing when you see them in the hallways can make a difference. Acknowledging their birthdays, rewarding high grades or improvement, simply greeting them and having non-content related conversation even for 5 seconds, can make a difference. Trying to get to know their home life, what they do for fun, any interests they have, their strengths and weaknesses both in and out of the math classroom can make a difference. What is important is that you show them that you're human and that you recognize that they're human as well. These are the things I learned in my undergraduate program that stuck with me and made my students recognize that I was different. I cared about their well-being, and it made them feel like they were respected. And that respect was almost always reciprocated. It doesn't matter how amazing your lesson plan is if you don't get to teach because the students are being disruptive. But when they know you care, they provide a wonderful environment for you to teach and for them to learn, making teaching mathematics a truly wonderful experience.

POSTSCRIPT

Alice F. Artzt and Frances R. Curcio

If we have learned nothing else from reading these touching and authentic untold stories, we have learned that great teachers are great caring human beings who have multiple abilities and talents and also have a passion for teaching, mathematics, and are dedicated to their profession and helping their students. We owe it to the future generations of students to provide teacher preparation programs that nurture and support preservice teachers so that they too can develop into the types of educators whose stories are just as inspirational as the stories in this book.

The teacher-authors all persevered through multiple challenges partly by relying on what they learned in their undergraduate teacher preparation program, as well as on the network of people who went through the program with them. At no point did any of these teachers feel alone or at a complete loss for how to think about the difficult times they faced, and, they all faced difficult times. Becoming a great teacher who can weather treacherous storms is no different from building a great building that can weather treacherous storms. Great teachers need to have a firm foundation of a teacher preparation program that not only focuses on instructional practice, but that also provides a network of people with whom they develop a close relationship and can call on at all times. They need a teacher preparation program that has instructors who are available and understanding of them as they will be available and understanding of their own students. They need a teacher preparation program that places them in teaching and leadership positions, long before they enter the classroom as professionals. They need a teacher preparation program that will help them utilize their unique talents in ways that

The Inspirational Untold Stories of Secondary Mathematics Teachers, pages 89–90.
Copyright © 2020 by Information Age Publishing
89

will enhance their teaching and be advantageous to their students. They need a program that directly exposes them to the professional teaching organizations so that they can participate in these organizations once they do become professionals. Such a teacher preparation program cannot occur within a summer stint or a few short semesters. Recruiting students directly from high school and working with them in a concentrated, coherent manner for four years is what creates the strong bonds and depth of understanding of teaching and learning that the authors of these chapters have demonstrated. As it is often said, teaching is a noble profession. The preparation of teachers must be ever more thoughtful if it is to prepare teachers worthy of such a distinguished role in society.

REFERENCES

Artzt, A. F., & Curcio, F. R. (2007a). Reforming mathematics teacher preparation: Now is the TIME!" In V. V. Orlov (Ed.), *Issues in theory and practice in the teaching of mathematics* (pp. 72–81). St. Petersburg, Russia: Russian State Pedagogical University Press.

Artzt, A. F., & Curcio, F. R. (2007b). TIME 2000: A mathematics teaching program. *Mathematics Teacher, 100*, 542–543.

Artzt, A. F., & Curcio, F. R. (2008). Recruiting and retaining secondary mathematics teachers: Lessons learned from an innovative, four-year, undergraduate program. *Journal of Mathematics Teacher Education, 11*(3), 243–251.

Artzt, A. F., Curcio, F. R., & Sultan, A. (2013, April). Queens College: A program for math teachers requires a complex formula. *Phi Delta Kappan, 94*(7), 23.

Artzt, A. F., Curcio, F. R., with Weinman, N. (2005). Hosting a conference for high school students: An innovative recruitment strategy. *The Urban Scholar, 2*(1), 4–5.

Artzt, A. F., Sultan, A., Curcio, F. R., & Gurl, T. (2012). A capstone mathematics course for prospective secondary mathematics teachers. *Journal of Mathematics Teacher Education, 15*, 251–262.

Curcio, F. R., & Artzt, A. F., with Porter, M. (2005). Providing meaningful fieldwork for preservice mathematics teachers: A college-school collaboration. *Mathematics Teacher, 98*, 604–609.

Einstein, A. (1935, 4 May). Letter to the editor: In appreciation of a fellow mathematician. *The New York Times*, 12.

Eiseley, L. (1964). The star thrower. In L. Eiseley, *The unexpected universe* (pp. 67–92). New York: Harcourt, Brace, & World, Inc.

The Inspirational Untold Stories of Secondary Mathematics Teachers, pages 91–92.
Copyright © 2020 by Information Age Publishing

Fang, Y., Nandy, A. M., Liew, D., Sun, H. M., Tan, S. P., & Khan, S. (2016). *Singapore teachers: Narratives of care, hope and commitment.* Singapore: World Scientific Publishing Co., Ptc. Ltd.

Ingersoll, R., & Perda, D. (2009). *The mathematics and science teacher shortage: Fact and myth.* CPRE Research Reports. Retrieved from: https://repository.upenn.edu/cpre_researchreports/51

Kierkegaard, S. (2015/1835). Journal AA: 12, Gilleleje,1 Aug 1835. In K. B. Söderquist, & B. H. Kirmmse, (Eds.), *Kierkegaard's journals and notebooks* (vol. 1, p. 19, A. Hannay, Trans, p. 19). Princeton, NJ: Princeton University Press.

Mulvahill, E. (14 June, 2019). Why teachers quit. *We are teachers.* Retrieved from: https://www.weareteachers.com/why-teachers-quit/

National Academy of Sciences. (2006). *Rising above the gathering storm.* Washington, DC: National Academies Press.

National Research Council. (2002). *Learning and understanding: Improving advanced study of mathematics and science in U.S. schools.* Washington, DC: National Academies Press.

Rising, G. (2019). *Letters to a young classroom teacher.* Hershey, PA: William R. Parks Publishing Co.

Scharton, H. (2018). Taking on teacher attrition. *eSchool* News: Today's *Innovations in Education.* Retrieved from: https://www.eschoolnews.com/2018/02/20/taking-teacher-attrition/

Tiny teaching stories. (2019). Retrieved from: https://www.edweek.org/ew/articles/2019/10/14/tiny-teaching-stories-for-good-instead-of.html?cmp=eml-enl-eunews1&M=58955887 &U= 12557&UUID=27ca4bd717c4786146abd70018ef813d

Walker, T. (3 April 2019). *Teacher shortage is 'real and growing, and worse than we thought.'* neaToday. Retrieved from: neatoday.org/2019/04/03/how-bad-is-the-teacher-shortage/?utm_s

U. S. Department of Education. (2002). *Meeting the highly qualified teachers challenge: The secretary's annual report on teacher quality.* Washington, DC: Author.

ABOUT THE AUTHORS

Kendal Askins

A native of Queens, New York, Kendal graduated from Queens College in the TIME 2000 Program in 2007. In 2012, she completed a master's degree in secondary mathematics education also at Queens College. In partial fulfillment of the requirements, Kendal wrote a thesis entitled, *Integrating Reflective Writing in Eighth-grade Mathematics Instruction for English Language Learners.* Kendal holds an advanced certificate in educational leadership from Wilkes University. After teaching in middle schools in New York City and in the East Stroudsburg School District, PA, Kendal is presently an assistant principal at Pleasant Valley Intermediate School, Kunkletown, PA.

Alexandria Capozzoli

Alexandria was born in Huntington, New York, and she graduated from Queens College in the TIME 2000 Program in 2010. She continued to study at Queens College, earning a master's degree in secondary mathematics education in 2013. The title of her thesis was, *The Impact of Implementing a School-wide Positive Behavior Interventions and Supports Program on Student Achievement in Mathematics in an Urban Middle School Setting.* From 2012 to 2016, Alexandria was a master teacher for Math for America. As a middle school mathematics teacher at IS 204, Long Island City, New York, Alexandria served as a lead teacher for the

The Inspirational Untold Stories of Secondary Mathematics Teachers, pages 93–96.
Copyright © 2020 by Information Age Publishing
93

mathematics department, and math lead teacher/math coach. At the time of this publication, Alexandria was on childcare leave.

Daniel De Sousa

Born in Bellerose, New York, Daniel graduated from Queens College in the TIME 2000 Program in 2017. At the time of this publication, he was teaching at Business Technology Early College High School, Queens Village, for three years. A candidate for a master's degree in secondary mathematics education at Queens College, he is also the mathematics content leader in the high school.

Angelina Ezratty

A native New Yorker, Angelina was born in New Hyde Park, and raised in College Point. She graduated from Queens College in the TIME 2000 Program in 2017, and at the time of this publication she was in her third year teaching seventh-grade mathematics at I.S. 125 Thomas J. McCann Woodside Intermediate School, Woodside, NY. She completed a master's degree in Secondary Mathematics, Grades 7–12, at Queens College. The title of her thesis was *The Effects of Incorporating Writing in Mathematics on Lower-Achieving 7th-Grade Students' Conceptual Understanding*.

Sabrina Joseph

Sabrina was born in the Bronx, New York. She graduated from Queens College in the TIME 2000 Program, in 2010. She earned a master's degree in 2013, from The City College of New York. Her thesis was entitled, *The Effects of Problem-Based Instruction on Student Confidence and Perseverance*. She earned an advanced certificate in school leadership from Massachusetts College of Liberal Arts. After secondary teaching mathematics for 9 years, most recently at John Bowne High School, Flushing, New York, Sabrina is an assistant principal at Pan American International High School, Queens, New York, fulfilling her dream to maximize her support of students, making a difference in their lives.

Young Mee Kim

Born in Seoul, South Korea, Young Mee graduated from Queens College in the TIME 2000 Program, in 2002. She remained at Queens College to earn a master's degree in secondary mathematics education in 2004. Her master's thesis was entitled, *Comparison of Metacognitive Cues in Open-ended Items on Selected New York State Mathematics A Regents Exams and on Selected Sequential Course I Regents exams*. Young Mee went on to earn a doctorate in education from Teachers College, Columbia University, in 2011. The title of her dissertation was, *Implementation of Cooperative Learning by Secondary School Mathematics Teachers*. For eight years, Young Mee has been teaching at Vaughn College of

Aeronautics and Technology, Flushing, New York, where she is currently an associate professor of mathematics.

Irina Kimyagarov

Born in Uzbekistan, Irina graduated from Queens College in the TIME 2000 Program in 2003, and completed a master's degree in secondary mathematics education in 2006. The title of her thesis was *The Effect of Formal Instruction in a Statistics Unit on the Graph-based Data Analysis Skills of Eleventh-grade Mathematics B Students.* In 2009, she earned an advanced certificate in educational leadership at Queens College. Since 2003, Irina has served as mathematics teacher at Elmont Memorial High School, Elmont, NY. For the past eleven years she has also served at the mathematics department chairperson, working with in-service and preservice secondary mathematics teachers.

Maria Leon Chu

Born in Trujillo, Peru, Maria graduated from Queens College in the TIME 2000 Program in 2010. She earned a Master of Science degree in Secondary Mathematics Education from Queens College in 2013, co-authoring a thesis entitled, *Selected Secondary Mathematics Teachers' Passion toward Teaching and Their Potential to Remain in the Profession.* In addition to being certified to teach secondary mathematics, Maria also holds New York State Certification in Bilingual Education—Spanish. In 2013, Maria began an Early Career Fellowship with Math for America, and became a Master Teacher Fellow in 2017. At the time of this publication, Maria was in her eighth year teaching grades 10–12, at Francis Lewis High School, Fresh Meadows, New York.

Michael London

Born in Flushing, New York, Michael graduated from Queens College in the TIME 2000 Program in 2007. As a requirement for a master's degree in secondary mathematics education at Queens College, he wrote a thesis entitled, *The Effect of Conceptual and Procedural Questions on Homework Assignments on Students' Performance on a Population-growth Exam,* completing his degree in 2009. Michael has been on the faculty of The Queens School of Inquiry, teaching in grades 9 through 12, since 2010, and he is an adjunct lecturer in the Mathematics Department at Queens College. From September 2015 to June 2019, he was a master teacher for Math for America.

Mara P. Markinson

Mara is a member of the Queens College faculty, teaching courses in the Mathematics Department and the Department of Secondary Education and Youth Services. Born in Manhasset, NY, Mara graduated from Queens College in the

TIME 2000 Program in 2012, and was on the faculty of the East-West School of International Studies, teaching grades 9–12, from September 2012 to June 2018, and served as the mathematics department chair from 2015–2018. Since 2016 she has been a mathematics instructional coach for the New York City Department of Education. From 2014–2018, Mara was an Early Career Fellow and Single-Session Workshop Facilitator for Math for America. In 2015, Mara earned a master's degree from Queens College, and wrote a thesis entitled, *The Effects of Culturally Responsive Classroom Management on Urban High School Algebra Students' Motivation and Confidence: A Self-Study.* In addition to holding New York State Certification in Mathematics, 7–12, Mara has also earned a Certificate of Completion: Effective Mentoring and Coaching of Novice Teachers, from New Visions for Public Schools. She earned a Master of Philosophy degree from Teachers College, Columbia University, where she is completing her doctoral dissertation entitled, *The Roles of Textbooks and Pre-Service and Novice Teachers' Knowledge and Beliefs in the Teaching and Learning of Geometric Proof.*

Nerline Payen

Born in Queens, New York, Nerline graduated from Queens College in the TIME 2000 Program in 2014. She continued in the master's degree program in secondary mathematics, completing her requirements and graduating in 2017. The title of her co-authored thesis was, *The Influence of Parent Education on Secondary Students' Achievement and Their Perception of the Importance of Mathematics.* Nerline is fulfilling her dream of reaching all learners with mathematics as she teaches sixth and seventh grades at George L. Ryan Middle School MS 216Q, Queens, New York.

Julio Penagos

Born in Medellín, Colombia, Julio graduated from Queens College in the TIME 2000 Program in 2006. He continued his studies at Queens College, earning a master's degree in secondary mathematics education, in 2014. To fulfill a degree requirement, Julio conducted an action research project and wrote a thesis entitled, *The Process of Negotiation of Meaning and its Effects on Motivation in the Mathematics Classroom.* Julio taught mathematics at Foundations Academy, Brooklyn, New York, and more recently, in an early college high school, The Queens School of Inquiry, Flushing, New York. Julio served as an adjunct lecturer in the Mathematics Department at Queens College, and was a Math for America fellow. At the time of this publication, Julio was an educational consultant, helping to improve the teaching and learning of mathematics throughout the northeast region.

SUBJECT INDEX

A
advanced methods, 10
advice, ix, 7, 9, 10, 48, 67
anecdote(s), 28, 29
anxiety, 16, 17, 21, 22, 32, 38, 39, 46

B
behaviorist strategies, 25

C
caring, 2, 7, 29, 81ff.
classroom management, 7, 22, 29, 88, 96
coach, 7, 35, 41, 96, 112
community, xii, xv, xvi, 4, 5, 40, 55, 56,
    57, 63, 64, 68, 75
conference, xv, xvi, 32–34, 36–38, 40, 41,
    43, 55, 56, 64, 68, 71, 72, 79, 86, 91
confidence, 4, 13, 33, 40, 42–44, 55, 65,
    94, 96
contracts, 7
cooperative learning groups, 7, 69, 70
coursework, xii, xiii ff., 32, 43, 78
cry, 10, 11, 17, 29, 47

D
demo lesson, 9, 19
difficult home life, 27
diverse cultural backgrounds, 11
divine intervention, 14, 15
doctoral degree, 41, 68
dream(s), xi, 14, 26, 33, 47, 48, 50ff., 61,
    62, 68, 80, 94, 96

E
economic disparity, 46
embarrassment, 31
encouragement, 6, 10, 70
enthusiasm, 21, 27
excitement, 2, 18, 27, 41, 75, 85

F
failure, 10, 22, 30, 54
family, xv, xvii, 5, 11, 13, 14, 17, 26, 27,
    46, 48, 50, 53, 55–57, 60–62, 68, 79, 86
fear, x, 4, 15, 41, 51
field observations, fieldwork, 25, 27–30
financial aid, 48
fractals, 74
friendship(s), xiii, 46, 49, 65

The Inspirational Untold Stories of Secondary Mathematics Teachers, pages 97–99.
Copyright © 2020 by Information Age Publishing
97

NAME INDEX

A
Absent Teacher Reserve, 8
Artzt, A., xii, xiii, xiv, 91
Association of Mathematics Teachers of
    New York State (AMTNYS), 38

B
Borges, J. L., 47

C
Celebrating Mathematics Teaching, 32, 36
Child Protective Services, 88
Children First Network, 85
China, 43, 49
Colombia, 45, 47, 96
Common Core, 11
Core Inquiry, 21, 22
Curcio, F., xii, xiii, xiv, 91

D
District 75, 7

E
Einstein, A., 50, 91
Eiseley, L., 85, 86, 91

English Language Learners (ELLs), 20,
    72, 93

F
Fang, Y. et al., ix, 92

G
General Equivalency Degree (GED), 4
Gurl, T., xiv, 91

I
Ingersoll, R., xi, 92
Initial Clinical Experience, 35

K
Kierkegaard, S., 46, 47, 52, 92

L
LIMAÇON, 40
Long Island, NY, 14, 32

M
Marquez, G. G., 47
Medellín, Colombia, 45, 96
MoMath Museum, 75
Mulvahill, E., xi, 92

The Inspirational Untold Stories of Secondary Mathematics Teachers, pages 101–102.
Copyright © 2020 by Information Age Publishing